STANDING UP TO
CHINA

STANDING UP TO CHINA

How a Whistleblower Risked
Everything for His Country

Ashley Yablon, Esq.

BROWN BOOKS
PUBLISHING GROUP

Standing Up To China
How a Whistleblower Risked Everything For His Country

Brown Books Publishing Group
Dallas, TX / New York, NY
www.BrownBooks.com
(972) 381-0009

A New Era in Publishing®

Publisher's Cataloging-In-Publication Data

Names: Yablon, Ashley, author.
Title: Standing up to China : how a whistleblower risked everything for his country / Ashley Yablon, Esq.
Description: Dallas, TX ; New York, NY : Brown Books Publishing Group, [2022]
Identifiers: ISBN 9781612545585 (hardcover)
Subjects: LCSH: Yablon, Ashley. | Whistle blowers--United States--Biography. | Lawyers--United States--Biography. | Zhong xing tong xun (Corporation)--Corrupt practices. | National security--United States. | Organized crime--China. | LCGFT: Autobiographies.
Classification: LCC JF1525.W45 Y33 2022 | DDC 327.12730092--dc23

ISBN 978-1-61254-558-5
LCCN 2021921835

Printed in the United States
10 9 8 7 6 5 4 3 2 1

For more information or to contact the author, please go to www.AshleyYablon.com.

To Donna,
for your tireless support and the countless
sacrifices you made along the way.

This book is also dedicated to anyone who has
ever faced a challenge, however big or small, that
made them question who they are, what they are
made of, and, most importantly, what they would
sacrifice to do what is right. To those who chose to
face the consequences and live with them
—you are not alone.

Contents

Acknowledgments

I could have never written this book without the love and support of all my family and friends. To my mother Carolyn who has always made me her priority. To my brother Aaron who is my hero, ultimate supporter, and best friend. I am the luckiest person in the world to have you both as my family. To all my friends who lent me a shoulder to lean on throughout the process, I can't thank you enough. If I learned anything, it's that you can never adequately measure the true blessing of family and friends. To all my attorneys who believed in me and helped with your wisdom and your support. To my dear friend Dennis and your entire team, and to Milli and all the great people at Brown Books. I never thought we would get to the end of this journey. So, cheers to us all. We did it.

CHAPTER 1

Hit Refresh

July 12, 2012, approximately 4:07 p.m.

Donna and I lie on the floor of our bedroom. Our eyes are riveted to the laptop sitting open before us on the rug. We can't see or think of anything else. That eleven-inch screen has become our universe.

I'm hitting the Refresh button every five seconds—like a crazed lab monkey trying to score a food pellet from an empty dispenser. The only sound I can hear is my heart pounding in my chest and the methodical clack of the mouse button as my finger hammers it over and over.

Clack. Hit Refresh. Clack. Hit Refresh. Clack. Hit Refresh.

Why are we on the floor? Because this way we can't fall any further when the news hits. We're as low as we can go.

Clack. Hit Refresh . . .

Sweat is pouring off my back and running down my sides in rivulets. Not from the Dallas summer heat, which is in full boil. Not from the hot breath of Gable, our hundred-pound Bernese Mountain Dog, who is hovering over us, honing in on our anxiety level and trying to offer us comfort—but only making it worse.

No, the sweat is from panic. Pure, blind, animal panic. If the story we're anticipating hits the Internet as promised, our lives will never be the same. No more dream job for me. No more career. Maybe no more pulse.

Seriously. Donna and I have been told we could end up dead.

Clack. Hit Refresh . . .

What is the website we're watching as if it's the jury room door after a capital murder case? The Smoking Gun. The site has promised to break a massive story with global implications. And I'm at the dead center of the story. I'm the whole reason the story exists. And some very powerful people are not going to be happy with me. Not at all.

Hit Refresh . . .

How did I find myself in such an insane position? *I ask myself for the thousandth time. It was only nine months ago that I took my dream job as General Counsel at ZTE USA, a major subsidiary of one of the largest telecom equipment manufacturers in the world. Only nine months ago that I thought I had the world by the tail. Only nine months ago that Donna and I embarked on what we thought was the most exciting chapter of our lives.*

And now it's all about to come crashing down in spectacular fashion. All because a document was leaked. A confidential document that should have been kept under government seal and should never have seen the light of day.

Hit Refresh . . .

Maybe the article won't be published. Maybe the editors have come to their senses. Maybe my lawyers got through to them and talked them out of it.

Hit Refresh . . .

Maybe if they do print the story, it'll just be a tiny piece that no one really notices.

Hit Refresh . . .

Maybe the reporter will grow a conscience and change his mind.

Hit Refresh . . .

Or maybe ZTE will be fine with the whole thing. Maybe the company will just accept responsibility for its actions, wish me good luck, and turn a new page in its corporate history.

Hit Refresh . . .

Yes, and maybe a snowball has a fighting chance in hell.

Hit Refresh . . .

Hit Refresh . . .

Hit Refresh . . .

July 6, 2012 (six days earlier)

I turned left into the parking lot of ZTE USA in Richardson, Texas, feeling jittery and nervous in that waiting-for-the-other-shoe-to-drop way. Those feelings had become my norm in recent weeks. Ever since that fateful day in May.

As I steered my way toward the massive grey box of a building that ZTE USA called home, panic seized me by the throat. Several of my ZTE USA coworkers were standing around outside the building, confusion on their faces. My pulse began to race. What was going on here?

I eased my foot down on the brake and cruised toward my parking space. Why was everyone milling about outdoors like this? It was after nine. Why weren't they upstairs working?

The only explanation my mind could come up with was today must be The Day.

I'd been thinking about this day for weeks—stewing on it, dreading it, bracing for it—but now I felt completely unprepared for what was about to happen. How should I play this? What should my demeanor be? Should I just act casual? Pretend I have no idea what's going on? Yes. That was my only play here.

Get it together, Ashley. Get it together.

I shut off the car, feeling the heat of the day instantly pressing in, and sat still for a moment. Taking several deep breaths, I tried to draw calmness into my body. I checked myself in the rearview mirror, fixed my hair, straightened my tie, smiled at myself. My smile didn't reach my eyes.

I tried the smile again—better this time—then reached for the door handle. Might as well get this over with. Take the plunge. With one hand holding my briefcase and the other hand wiping sweat from my brow, I exited the car, lugging what felt like concrete shoes toward the building. My mind was racing ahead to worst-case scenarios.

The reason for my dread? About two months earlier, after weeks of agonizing and soul-searching with my wife, Donna, I had made the hardest decision of my life. I had decided to turn informer against my multibillion-dollar multinational corporate employer, ZTE. I had taken this dizzying step because I'd discovered that the mega-corporation I was working for was engaged in billions of dollars' worth of systematic illegal activity, selling US-made components to countries where such trade was banned—Iran in particular.

And now I couldn't "unknow" what I had come to know.

So on May 2, 2012, I had sat down with two FBI agents and said, "Okay, here I am. What do you want to know?" And over two long sessions, I had spilled the billion-dollar beans. I had spelled out exactly how I'd stumbled upon ZTE's illegal scheme. I had described to the agents, in detail, how a series of shell corporations had been created to disguise the company's shady activities. I had listed, by name, all the players involved, in both the United States and China. And the FBI had captured my story in a detailed, thirty-two-page affidavit.

The purpose of that affidavit, as explained to me, was, among other things, to obtain a warrant for the FBI to conduct an evidence-gathering raid on the US offices of ZTE.

And now, it seemed, the day of the raid was here.

In theory, I had nothing to panic about. My affidavit was under seal and was completely confidential. That meant—again, in theory—that no one at ZTE would know it was me who had blown the whistle, at least for the time being.

But we all know how theories sometimes go.

As I walked toward the building, I noticed something that didn't add up. One of my staffers was laughing with one of the company's top executives. What was so funny?

Then I spotted another incongruity. Many of the people standing around outside did not even work at ZTE. Why would they be standing outside during an FBI raid on ZTE? In fact, now that I thought about it, why would the ZTE people be standing outside? The FBI doesn't make you vacate the building when they conduct a search, do they? They keep all employees on hand to help them find the documents they're looking for, right?

I approached a woman in a suit who was taking cell-phone pictures of the impromptu gathering. "What's going on?" I asked her.

"Fire drill," she said, pointing to the fire marshall's car parked at the far end of the building.

I laughed. Relief poured over me like an ice-cold Gatorade bath. My paranoid brain had been jumping to conclusions. Today was not The Day. I felt lighter instantly, as if I'd shed a heavy coat. I headed toward the building with a new spring in my step.

My relief lasted about three seconds. Then my cell phone rang. The call was from a New York number I didn't recognize. I figured it must be related to one of the pending litigation cases I was handling for ZTE. I stepped off the walkway and tapped the Accept icon on my phone.

"Hello, this is Ashley," I announced.

"Mr. Yablon? My name is Mike Smith and I'm employed by Turner Broadcasting."

"Yes?"

"I work for a website called The Smoking Gun. Perhaps you've heard of us."

"I have."

"Then you know what we do. Mr. Yablon, I have come into possession of a copy of a thirty-two-page affidavit documenting a conversation you had with the FBI on May second."

"What?" My heart jumped like it was trying to exit my chest. Fear washed over me, erasing all sense of relief. I was sure I'd heard him wrong. I asked him to repeat himself.

He did. And then he told me, "I just wanted to inform you that I'm writing an article about the ZTE/Iran scandal, and it's going to be published next Thursday, along with the complete affidavit. I wondered if you had any comments you would like to add."

"You can't do that!" I shouted into the phone. "That affidavit was under seal! It's not public information."

"I'm afraid that will change come Thursday," he replied.

I pushed End Call and walked dizzily away from the building, hearing my heart hammering in my chest. This couldn't be happening. Couldn't.

I found a semi-private spot between two parked SUVs and hit the number for my criminal attorney, Ted Masters. (Yes, I now had a criminal attorney, as well as four other specialized lawyers, on my personal payroll—something I never could have imagined a few months earlier.) Waiting for Ted to answer, I nervously paced back and forth.

"Ted!" I shouted, the moment he picked up. "What the fuck is going on?"

"Take it easy, Ashley," he replied in his reassuring Texas drawl. "Why don't you tell me what's got you all riled up."

With as much calmness as I could muster, I explained the call I had just received. "How could this be happening?" I said to him. "If that story gets out, I feel I'm a dead man. I'm not kidding. I'm literally dead. I've heard stories about what happens to people like me. We're talking about billions of dollars' worth of business that will be lost because of my big mouth. Billions with a B. We've got to stop them from publishing that story and that affidavit!"

"And that's what I will try to do. I need to make some calls, see what I can figure out."

"How could this even have happened? I thought that affidavit was under lock and key in a federal courthouse. How could anyone have gotten hold of it?"

"I intend to try to find that out, my friend," said Ted. "You just need to calm down and leave it to me."

"It's illegal for anyone to have possession of a sealed affidavit, right? So that means it's illegal for anyone to publish it, right? Can't we get an injunction?"

My attorney was silent.

"Ted?"

"You're correct that the press is not allowed to obtain information illegally . . . "

"But?"

"But . . . if someone leaks information to a reporter, and the reporter hasn't done anything illegal to receive it, there's not a lot that can be done to stop it from getting out. Thanks to a little thing called the First Amendment."

As an attorney, I knew this in an abstract way, of course, but freedom of the press sure looks different when you're the one looking down the barrel of its gun.

"You've got to fix this, Ted!"

"I'll get on it right away," he said. "Just give me a little time to figure things out."

"In the meantime, what am I supposed to do?"

"Your job. Go to work. As usual. Keep quiet. Act normal."

"ACT NORMAL? HOW THE HELL AM I SUPPOSED TO DO THAT?"

I hung up and paced between the SUVs, my heart still pounding loud enough for me to hear it through the air. I knew who I needed to call next, but I was reluctant to do so. Donna was in Colorado at the moment, helping take care of her sister, who was battling breast cancer, and her three young boys. I knew that situation was extremely taxing on her emotionally. The last thing she needed to hear was bad news from the home front. But there was no keeping this from her.

I made the terrible call and described to her, with a shaky voice, the axe that was poised to fall on our necks in a few days' time.

"Stay calm, Ashley," she said, as if talking to one of her clients. Donna is a lawyer too. "I'll be on the next plane home."

June 12 again, approximately 5:14 p.m.

Now here we lie on our bedroom floor, drenched in sweat, watching that damn laptop like cats watching a mouse-hole, our hundred-pound dog's body-heat turning the room into a sauna.

Clack. *Hit Refresh . . .*

Still nothing on the website. It's after five o'clock. Maybe The Smoking Gun has gone home. Maybe the story is not going to be published today.

Clack. *Hit Refresh . . .*

What are we going to do if and when the story does hit? What's our next play?

Hit Refresh . . .

Hit Refresh . . .

Hit Refresh . . .

Six o'clock. Seven o'clock Eastern Time. They probably won't release the story now. Too late in the day; they won't want to miss the news cycle. Maybe we can relax—for today.

Hit Refresh . . . .

Uh-oh. This time, on about the four-thousandth click of the mouse button, there is a lag after I hit Refresh. Something new is loading. Oh shit. Oh no.

No, no, no.

Finally the headline splashes across the screen on a bright red, five-alarm background: "FBI Targets Chinese Firm Over Iran Deal." And right there, on page one, is my photograph—cropped from a photo of me proposing to Donna on Broadway years earlier. I speed-scan the story—my brain is too scrambled to read it in logical sequence—and see my name over and over again. The article is all about me and my role in uncovering the ZTE Iran scandal. It contains damning new information too, all of it attributed to me.

Donna jumps to her feet, and I do the same. She places her hands on my shoulders, looks me straight in the eye, and says, "Ashley, we have thirty minutes to get out of this house or we're dead."

She's right. Our home address is public knowledge. The press will be here within the hour, but they're the least of my worries. Others will probably be here sooner. It's the "others" that scare the shit out of me.

We grab a couple of duffel bags and run frantically through the house, grabbing whatever necessities we can think of—phones, charger cords, laptops, a few changes of clothes, some toiletries. We're out the door within minutes, along with our confused dog.

"Where the hell do we go now?" says Donna, tensed behind the wheel of our car.

"I have no goddamn idea," I reply. I'm wishing my brain had a Refresh button because right now it is stuck in an endless loop. The only thought that keeps running through my head, over and over, is How the hell did it come to this? How the hell did it come to this? How the hell did it come to this?

CHAPTER 2

An Actor Prepares

I'm Ashley Yablon, and I'm a whistleblower. In 2012, I discovered that my employer, ZTE, a huge multinational company with its home base in China, was deliberately and intentionally breaking US law, to the tune of hundreds of millions, if not billions, of dollars' worth of annual business.

I was faced with a dilemma: do I keep my mouth shut and continue to enjoy my nice corporate law career, or do I say something? The stakes were staggeringly high whichever way I chose. If I decided to shine a light on what was going on in the company, not only would the company lose billions of dollars in business but it would also face massive legal consequences from the US government. If I played along with the scam, then here was the ultimate picture I saw: me standing in front of Congress live on C-SPAN as the sole representative of ZTE USA, swearing to the lies of my Chinese employer. And with each false answer I gave, I'd be further committing treason because, despite my assurances otherwise, ZTE was indeed selling spying technology to embargoed countries against US laws. A federal prison cell likely awaited me.

Either way I chose to go, the entire status of US–China relations would probably be altered. No exaggeration. And it all came down to me.

I've never considered myself particularly heroic; I'll just put that on the table. I mean sure, I'm basically an ethical person, a good person—I am—but I'm no moral purist. As a lawyer, I've been known to look the other way when my clients or employers were flirting with the finer edges of the law. But the situation I stumbled on at ZTE was so egregious, so black and white, it essentially drew a line down the center of my conscience. And I knew whichever side of that line I decided to stand on was going to define me for the rest of my life. In that sense, the decision wasn't hard. But it was still the hardest thing I ever had to do.

The costs of my actual decision? Enormous. My employer did lose billions of dollars in business and endured the highest fine ever levied by a US court in history. And if you read the newspapers, you know ZTE is still staggering from the blows due to the huge restrictions that have been placed on its products and its market worldwide. ZTE may never fully recover.

My costs were huge too. I lost my dream job and my dream career. I lost my reputation in the job market where I earn my living—there was a period of over two years when nobody, literally nobody, would return my calls. I racked up attorney fees in the six figures. I lost a ton of money and had to borrow from family and friends. I lost relationships. I lost my pride, my faith in being protected as a US citizen, my health, and my peace of mind. I was pursued by dangerous people, and both Donna and I were threatened with death.

I eventually got most of the important stuff back. I'm still alive, and that's not nothing.

But my career? My finances? My relationships? My safety and security? Well, I'll tell you about those later in the book.

As I look back on the whole saga, I try to take some personal responsibility—or at least try to see how I got myself involved in this whole debacle. I ask, "What was my fatal flaw here? What did I do to get myself into this mess?" The simple answer is: naïveté, or something close to it. I put too much faith in people and situations. I saw only what I wanted to see. I wore rose-colored glasses. At the time I was hired by ZTE, I thought my dream job had been laid out before me, and so, without ever really weighing the potential risks or doing a deep dive into the company and circumstances, I jumped in head first. I bought into the myth I wanted to believe, and my belief blinded my perception.

I'm convinced, now, that I was hired to be the fall guy right from the start. I think ZTE was grooming me for that purpose from the moment

I walked in the door. But even before that, I think my life was grooming me to play my part in this drama. I was the perfect person to fit into the perfect role at the perfect time. Call it a perfect storm.

I won't give you my whole life story, but let me sketch in a few of the basics.

Dallas breeding

I'm a Dallas boy, born and raised. There were some rough edges around my childhood, but it was basically a good one, an optimistic one. During the '70s and '80s, my childhood years, my family lived in various Dallas neighborhoods. I rode my bike everywhere, built treehouses, and played any sport I could. When I was junior-high age, we moved to a new high-rise, the first of its kind in far North Dallas. It was surrounded at the time by horse stables, empty fields, and golf courses, a high-rise in the middle of what seemed like Mayberry.

Like any American family, we had our ups and downs. Regardless of our month-by-month financial situation, though, academics were always number one. No matter what grade, mark, or score I earned, for my dad it was always, "Oh, an A-. So you couldn't get the A+, huh?" Sometimes the remark was made with a tongue-in-cheek smile, but it always contained a measure of biting truth. That was the game we played, even through my second year of college, when Dad passed away.

My life wasn't problem-free by any means, but it was, for the most part, an un-conflicted one. The world I was raised in was one in which if you wanted to achieve a goal, you went after it. And if you put in the work, you got the reward. Expectations were high and goals were realized. My family wasn't wealthy, but we were more privileged than I realized at the time. I went to Jesuit College Preparatory School, one of the best-rated private prep schools in Dallas. Jesuit, if you haven't seen it, looks and feels more like a college campus—twenty-eight acres, football stadium, baseball field, art museum, state-of-the-art computer labs; you get the idea. I studied hard and was appropriately rewarded with good grades. I was on Student Council. I tried to embrace the Jesuit mantra of being "A Man for Others."

And so, when it came time to apply for colleges, I got into the good schools. I eventually went to Southern Methodist University (SMU) in Dallas and even earned a partial scholarship. But as I said, my parents weren't rich. They couldn't afford to foot the balance of my tuition, so—unlike many of my college friends—I had to work my way through undergrad school, especially now that my dad was gone. My peers wore t-shirts and shorts to class. I wore a jacket and tie because I had to go to work right afterward. I had a series of part-time jobs in college, and one of them changed my life.

Boy meets law

In my first year at SMU, Brad Keller, a family friend who was an attorney, got me a job as a runner at the Dallas law firm where he worked. In those days, most law firms hired college kids as runners; I guess they still do. "Runner" is a fitting name for the job. You literally run pieces of paper from one place to another. You also do light clerical work, clean out the lunchroom fridge, make coffee, and get the partners' cars detailed.

This was the early 1990s, so the computer revolution hadn't fully kicked in. A lot of what I did was make copies of cases for the lawyers from physical books (remember those?) on a Xerox machine and run legal documents to the courthouse. And, of course, fetch barbeque brisket sandwiches. It wasn't challenging work, but I liked it. The atmosphere agreed with me.

Brad was not only a friend but a mentor to me. I had idolized him since I was a kid. After I graduated cum laude from SMU, I stayed on full-time at the law firm as a runner. I wasn't sure what I wanted to do. Brad eventually talked me into considering law school. In my mid-twenties, I applied and was accepted at a couple of good law schools in the Northeast.

I had my plans all laid out to go to Ohio State—a top-fifty law school—when, in the winter of 1998, life threw me a happy curveball. I was introduced to Donna Yarborough. The fiancé of a friend of mine worked as a paralegal along with Donna, and these two mutual friends of ours wanted to set us up. Donna had decided to throw herself a party after taking the LSAT exam—that's the law-school equivalent of the SAT—and that's where I met her. I was blown away, to put it mildly. Donna was gorgeous. Piercing blue eyes, five feet tall, long brown hair. She was also fiercely intelligent, funny, irreverent, disarming, strongminded, and focused. You don't get in Donna's way when she's on a mission. I loved that about her. And her personality was—and is—extremely outgoing. She has never met a stranger, and she possesses that rare ability to make everyone from a CEO to a shoe shiner feel like her best friend.

Her favorite book was *Gone with the Wind,* and she liked to say, "Ashley, Ashley, Ashley!" in a perfect Scarlett O'Hara drawl. I fell for her on day one, and we started dating.

On our second date, I snuck us into a hotel ballroom and proceeded to mangle the cork on the bottle of wine I'd brought along. Donna jumped up and ran downstairs to the hotel bar to get a better corkscrew, and I remember thinking, *I love the way she just takes action!*

Donna was planning to go to law school at Loyola in New Orleans. Me? I wanted to go wherever she was going. University of Siberia? Perfect. Afghanistan State? Sign me up. I applied to Loyola at the last minute and was accepted there.

Loyola

Donna and I moved to New Orleans together in August '98 to attend Loyola. It had to be the hottest day of the month as we pulled up in our

U-Haul on day one. We had rented a 1920s "shotgun double" close to the law school campus—"double" because it was a whole twelve feet wide; "shotgun" because you could supposedly stand at the front door, fire a shotgun and watch the buckshot fly through the house without hitting anything on its way out the backdoor. All the rooms were lined up in a row, no halls. We loved that old house and spent many a weekend night there dancing in the dark—mainly because we owned nothing but a stereo and a collection of old CDs. Our favorite? Van Morrison's *Moondance*. I still know every line from every song on that album.

We experienced our first hurricane together (Hurricane Georges) and the desperation one feels fleeing a city. Donna wisely filled an ice chest with food and water—which we lived on as we sought refuge. While escaping New Orleans, the stop-and-go from the highway's traffic gridlock was causing my fifteen-year-old car to overheat. Despite the sweltering New Orleans temperatures, we were forced to turn the car's heater on full blast to draw heat off the engine in an attempt to cool it down. It worked, at least long enough to get us out of town. For the next ten hours, we labored on to Mississippi (which, under normal conditions, should only take three hours). Into the wee hours of the night, we stopped at every motel looking for a vacancy. In a small town whose name I can't recall, we finally got the last room available—due to a no-show—and snuck Donna's Dalmatian, Elliott, into the motel with us. Eventually we made our way to Starkville, Mississippi, where we stayed with Donna's sister.

That first year of law school, I was lucky enough to snag a part-time job working in downtown New Orleans for Senator John Breaux's office, where I gained some valuable political experience. After the first year of law school, law students typically get jobs or internships at law firms. Not I. I was on a different track. I was able to land a small, paying summer job at Senator Breaux's office in Washington, DC. My job took me all the way to Capitol Hill! That experience sparked some political ambition in me, at least for a while.

In year two at Loyola, I was the first second-year law student ever to serve as Vice President of the Student Bar Association. The election had been at the end of our first year. Donna helped me run my campaign. I won the election in a landslide, which was a huge honor and accomplishment. I also made the Moot Court team and ATLA, a mock trial team—as did Donna. Fittingly, she was my trial competition partner. We competed in high-profile events with other law-school programs—another massive honor.

The life plan was unfolding right on schedule: you sow the efforts, you reap the rewards. Even better, Donna and I were learning we made a great team in more ways than one. After two years at Loyola, though, we both decided we wanted to be back home, so we returned to Dallas and spent our final year of law school at SMU where we had both been undergrads, though we'd missed each other by being in different class years. With both of us taking on full-time course-loads and working nearly full-time jobs to pay our bills, we finished our third and final year at SMU.

After graduation, we buckled down together to study for the bar exam, eager to see what our legal futures held in store for each of us. In November of 2001, we both passed the bar. Same day, right in synch. It was onward and upward for the team of Yablon & Yarborough.

A glimpse of the grail

Not long after passing the bar, I got in touch with Brad again. He was no longer working at the law firm but had become General Counsel for a big corporation. I didn't know what a "General Counsel" did exactly, but when he explained the job—and the perks—to me, I felt a veil lifting from my eyes.

"I love it, Ashford," he told me (Ashford was his nickname for me since I was a little kid). "I get to do a whole range of interesting things. I work nine to five, I answer only to the board of directors, and I get a very nice, very steady paycheck, along with stock options, a retirement plan, and plenty of vacation time. I would never go back to working at a law firm."

I was sold. I wanted to do what Brad did. I wanted to be a General Counsel.

I had found my dream.

Me and about four trillion other attorneys, it turned out.

Working "in house," I soon discovered, was the dream job for most lawyers—and becoming General Counsel, or "GC", was the Holy Grail. What does a General Counsel do? He or she oversees and coordinates all the legal activity within a large to very-large company. As a GC, you work with the heads of all the various departments—sales, engineering, marketing, design, human resources, finance, whatever—to identify whatever legal issues they may be facing and to provide advice and direction. You also put out fires, manage outside counsel, and handle legal compliance issues, all while getting involved in things like taxes, M&A, intellectual property issues, and labor law. A little bit of everything.

The best way to describe the difference between being an attorney at a law firm and a GC attorney at a company? An attorney at a law firm specializes in one type of law but has many clients. A GC attorney at a company has only one client (the company) but handles many types of law—a jack-of-all-trades.

People covet the job because it is varied and interesting, but also because it's a damn sweet gig. You don't have to deal with all the law firm b.s. What most people don't realize about working at a law firm is that it pretty much sucks. If you're "lucky" enough to get hired by a good firm fresh out of law school, they work you to the bone, 'round the clock, for your first several years. You've got no life during this period; the firm owns you. Your dangling carrot is the hope of becoming an associate and getting on the partnership track. But to be considered for partner, you need to be constantly bringing in clients, which amounts to sales work and is a pain in the ass, not to put too fine a point on it. My legal buddies were all climbing law firm ladders, becoming experts in one type of law or another.

Not me. I wanted the GC gig. As in-house counsel, life is simpler and far more pleasant. You earn a very nice salary, you don't have to woo clients, and you work bankers' hours.

Which is why everyone wants the job. Which is why the job is almost impossible to get. To become General Counsel, you typically have to work your way, slowly, up the pyramid—from corporate counsel to associate to assistant General Counsel. And then, if you're good enough, and you stick around long enough, maybe you make General Counsel. Very competitive.

So I knew in-house work was going to be tough to get, but that didn't stop me from asking. I arranged to have lunch with Brad one day and asked him, point blank, "What are the chances of my getting in at your company?"

Brad leaned back in his chair and thought for a moment, and I'll never forget what he said: "I could hire you, Ashford, but the problem is, you don't know anything."

Oh, just that.

I think he meant his words in the kindest way, but still they stung.

"You need to round out your tool belt," he continued, "before you even think about looking for work as an in-house attorney."

"Rounding out the tool belt" was some of the most important advice I've ever received. I didn't realize it at the time but this single phrase set the tone for the rest of my life.

"What do you mean?" I asked.

"I mean you need to gain some knowledge and experience first. They call the job 'General Counsel' for a reason. You need to learn to be a generalist. Right now, most of your friends from law school are out there trying to specialize, trying to become experts within some very narrow legal space. You need to do the opposite."

"How so?"

"Go horizontal, not vertical. Instead of learning a lot about one small area of the law, learn a little about a bunch of different areas. A General Counsel needs to be able to talk and think intelligently across a whole range of legal issues and disciplines. If you're serious about doing this job, Ashford, you need to go learn about litigation, learn about commercial contracts, learn about employment law, patent law, tax law. Learn about compliance and human resources. See what I mean? Round out your tool belt."

CHAPTER 3

Getting to Work

So that's what I did. I spent the next six years "rounding out my tool belt." It was like going back to school again. While my colleagues were working their way up their specialized legal ladders, I was working at a series of entry-level jobs. My first one was as an associate attorney in a boutique litigation firm. I did trials and depositions—and hated it. There's nothing like hearing your boss telling you the suit, shirt, and tie combo (navy and pink) you wore to a hearing was "a joke," and that you "would have looked more appropriate in a g-string or a tutu." I guess he thought I was a tad overdressed for a courtroom in rural Texas—hence the nickname "Fancy" which my coworkers hung on me. To me, my suit was just my Uniform[1]. If I had a vice, it was that I liked nice clothes and dressing well.

In another job I learned about contracts, in another about human resources, in yet another about how to do transactional stuff. Each time a new opportunity came along, I grabbed it. At each of these jobs, of course, I had to take a cut in pay or at best make a lateral move rather than earn a raise. Why? Because I didn't bring any expertise or experience to the table. I was a newbie. I was there to learn. In fact, during those years, I remember being so broke with each new job, Donna would bring me a weekly bag of groceries so I'd have something to eat at work. Rounding out my tool belt was putting me financially behind all my law-school friends. As they

---

1. I had something of a "uniform" for work. It consisted of a dark suit, white or striped shirt, French cuffs, and solid or striped tie. White linen pocket square. And nice dress shoes. My suit pants were cut with no break on the hem or cuff. And no socks. Never ever socks. Can't do socks. I've been told this is my look. My Uniform. Never thought about it—probably because I have been wearing a coat and tie since my Jesuit days—but I guess others do.

continued to specialize and earn more, I continued to generalize and earn less.

Nonetheless, I was focused and committed in my career, and so was Donna. We got married on October 17, 2003, a Friday, and we didn't even take a honeymoon. In fact, we spent the morning of our wedding at Donna's office going through documents to help her boss prepare for a hearing he was handling that day. We were both back at work on Monday morning. Donna was working as a trial attorney, which she loved, and I was busily grooming myself for that in-house job I hoped would be coming my way.

I did, however, give Donna a memorable marriage proposal in 2003, which was captured in photos as the winning entry in Korbel Champagne's Perfect Proposal contest. I had seen an ad in GQ magazine asking guys to send in their perfect proposal idea. If you were one of the three finalists chosen, your proposal idea was brought to life, complete with a diamond engagement ring. The ultimate winner received $10,000 for their wedding and honeymoon. What a deal for a broke young lawyer! My idea? Sing on a Broadway stage, in New York City, in front of a full crowd and "sweep Donna off her feet." I won the contest by singing at The Palace theater on the stage of the Elton John musical *Aida*. Our proposal photo found its way back to the front page of *The Dallas Morning News*.

A path opens

Finally, one day in May 2006, I got a call from my friend Robert Winston who worked at McAfee, the anti-virus software company. They had an office in Plano, Texas, just a few tee-shots up the road from Dallas. Robert explained that an incident had recently occurred at the company which had caused a mass exodus of in-house legal staff. He then gave me the news I'd been waiting and hoping for: They needed a jack-of-all-trades for their in-house legal department, pronto. Well, call me Jack. I'd spent the last six years studiously becoming a non-expert in multiple legal disciplines in preparation for just this opportunity.

They hired me. My career as an in-house attorney had finally begun. I knew this was a big opportunity, and I was determined not to waste it.

It was an interesting time to come aboard at McAfee. The former GC, Kent Roberts, had recently been fired amidst accusations of improper stock transactions, and prosecutors were speaking publicly about going after him personally. This served as a chilling notice to me that even as an in-house attorney you are not immune from personal indictment. On the other hand, news also came out about Roberts' pay. He was earning nearly $600,000 a year in salary and bonuses and had recently sold his company stock to the tune of nearly $7 million.

That was the club I wanted to belong to.

I had just barely started at McAfee when a memorable meeting took place. The new GC, a man by the name of Matt Caldwell, sat us down together in a room—all of the in-house attorneys and paralegals—and said, "I want you all to know: We will always do the right thing in this office. That will be always our barometer."

Those words imprinted themselves on my brain. As a lawyer, it is truly a gift to know that your employer has a moral compass. That way, you know you're never going to be asked to do something that goes against your conscience or to cover for the sleazy behavior of others. Matt Caldwell became a model to me—living proof that one can be both a corporate attorney and an ethical human being. I wanted to be like him. By the way, eventually, Kent Roberts was cleared for any potential wrongdoing.

I jumped in with both feet at McAfee. I started by helping handle the procurement contracts. These are contracts for the everyday things that run the company, such as the computers, the Xerox machine, the employee insurance plans, et cetera. I also helped with a companywide e-discovery system. (When a company gets sued and the other side asks for documents, an e-discovery tool goes through every employee's email and pulls out relevant docs.) I assisted in compliance and HR matters too, redrafting and revising forms and policies. I made a good impression with my bosses. My generalist skills turned out to be just what the department needed; Brad had given me great advice in that regard.

For my part, I was getting a chance to learn the ropes and politics of being a good in-house attorney. I learned a lot of excellent lessons at McAfee, lessons that would shape my thinking about the job of in-house counsel.

The stars were in perfect alignment, and I was doing everything I needed for my career to be a successful one.

I stayed at McAfee for about four years, working my way up from Corporate Counsel to Senior Corporate Counsel. One day, a friend I'd made at McAfee gave me a call. He had now moved on to Huawei, the megabillion-dollar Chinese telecom manufacturing company. Huawei had a US headquarters in Plano, practically next door to McAfee.

This friend presented me with what seemed like the perfect next step toward my dream. In May 2010 I was offered a job at Huawei USA. It was the equivalent of an assistant General Counsel position. I would be one tier away from the top of the pyramid.

The stars were continuing to align.

The China connection

Huawei is a massive corporation. It makes smartphones and other consumer electronics, as you probably know, as well as a wide range of networking equipment. At the time I came aboard, Huawei was a Global Fortune 500 company about to unseat Ericsson as the largest telecommunications manufacturer in the world. It had tendrils in over 150 countries and was providing products and services for forty-five of the top fifty telecom companies in the world.

As you may also know, Huawei has recently run into serious trouble in the US, partially due to US concerns about Huawei's technology enabling the Chinese government to spy on the US and other countries, and partially because the company has been implicated in the sale of information technology to Iran and other countries in violation of US sanctions. Both of these issues are central to this story, and we'll come back to them very soon.

For me, getting a job at the US headquarters of Huawei (Huawei USA) was the ideal move at the ideal time. The company was at the forefront of one of the most exciting and important industries in the world, and working there would give me an opportunity to learn about the inner world of Chinese multinational corporations. Unless you've had your head under a rock for the past decade or two, you probably know that China is the new eight-hundred-pound gorilla in the world economy. Today's global corporate game is you either beat China (for now), you learn to work with her, or you get out of her way. China is the new wave sweeping the planet. I knew that if I wanted to have a General Counsel career in today's corporate world, I'd better get up to speed on China. And what better way to do that than from inside one of her biggest companies? So the Huawei opportunity looked like a golden one. As with McAfee, I jumped in with both feet.

My job put me in a unique position in relation to Huawei's other legal team members. I needed to learn about Chinese business culture and psychology from them; they needed to learn about US law from me. Both sides, it turned out, had a lot to learn.

My boss, the General Counsel, was a woman named Heather. Of course, Heather wasn't her real name. Like many executives in Chinese American companies, she used an Americanized version of her Chinese name when working. The Chinese didn't do this so much out of respect for Americans as out of an opinion that Americans were too stupid and lazy to learn their real Chinese names (a justifiable opinion, if I'm being totally honest). This slightly superior-minded, bemused tolerance of Americans was an attitude I would find lurking behind the polite smiles of almost all the Chinese executives with whom I would do business.

I quickly came to see that there really is a difference between the Chinese mindset and that of most of the rest of the modern Western world. This difference no doubt stems from the fact that for so much of its history, China evolved along its own track, largely isolated from the West. And when the Chinese did have dealings with the West, things did not necessarily go well for China. So there is a deep-seated distrust of foreigners embedded in Chinese business culture. Ultimately, of course, people are people everywhere, and they have many more things in common than things that divide them. Still, Chinese culture, particularly Chinese business culture, is different from ours, and if you fail to learn that truth, you fail at your own peril.

At Huawei, I tried to be a sponge and learn as fast as I could. Naturally, I had to learn the protocols around business and social interactions, such as exchanging gifts, entering and leaving rooms for meetings, sitting for meals, and exchanging business cards—as well as key phrases and gestures in Chinese. I also had to learn what was acceptable and unacceptable professional behavior; for example, that physical touching and expressive body language were frowned upon at meetings, while excellent preparation, high-quality paperwork, and respectful posture were encouraged. That was the easy part.

Harder to learn were the subtler things. For example, in Chinese culture there is a strong emphasis on "face"—as in "saving face." The concept of face is similar to reputation, and virtually all business interactions are seen as opportunities to either build face or lose face. There are always face considerations at play in business dealings, on a meta level, and if you fail to understand that, you will miss a lot of what is really going on. Building face is seen as an incremental process: you do it slowly, over time. That is part of the reason deals can often take a frustratingly long time to cement. There is a courtship period in which face must be built.

Losing face is considered very bad and needs to be avoided at all costs. Causing someone to lose face—by blaming or finding fault with them in any direct way—is also disastrous and can ruin multimillion-dollar business deals. Because of this, people jump through hoops to avoid putting others in a position where they will lose face. This leads to a culture that feels to Westerners like indirectness and a failure to openly state what is on one's mind.

This seeming lack of directness is furthered by the fact that Chinese business culture is not as linear, black-and-white, and logically driven as Western culture. Americans tend to reason like, "A + B = C, so let's do C. Great, done!" In Chinese culture there is often a more circuitous route to conclusions. The path is more spiral-shaped. Many different options and considerations are weighed, in a more open-ended way.

For these reasons and more, communications with Chinese businesspeople can be tricky for Americans to master. It is generally considered rude, for example, to simply say no or to be confrontational in one's speech. You don't challenge another person's version of things. Very often when a Chinese businessperson says, "Don't worry about it," or "It's not a problem," they mean the exact opposite. But they can't come out and say so. As an American, you typically feel—often rightly so—that there is a level of communication going on beyond the words, and you don't understand everything that's really being said.

This air of ambiguity, I was shocked to learn, also applied to matters of law. We Americans—especially we American lawyers—tend to see the law as cut-and-dried. Once you understand the law pertaining to a particular field of action, you pretty much follow it. If you choose to disregard it, you know you are committing a crime or misdemeanor. Chinese business executives, I was discovering, don't see things quite this way.

I spent a lot of time trying to explain US law to my Chinese colleagues. There were many conceptual and linguistic hurdles to overcome in this regard. I found myself, on several occasions, literally drawing pictures on a whiteboard in an attempt to communicate some key aspect of the law. It wasn't that my colleagues were stupid; far from it. Rather, there was a fundamental disconnect between us. They seemed maddeningly unwilling to understand the simple concept: "You can't do that because the law forbids it."

One way this disconnect came to my attention, for example, was through the company's use of visas. Most employees at the Huawei facility—I'd say about eighty percent—were Chinese nationals, here in the US on visas. Almost everyone in the company had an H-1B visa. These visas are supposed to be issued to people with unique skills, in order that a company may fill specialty roles that would be difficult to fill with native US citizens. But at Huawei, everyone had H-1Bs—secretaries, project managers, maintenance people. The fact that such a huge, established company would institutionally disregard, or deliberately misinterpret, US immigration law seemed a bit shocking to me. (Huawei was actually busted by ICE for this practice while I was working there.) Boy, did I have a lot to learn.

The Chinese attitude toward US law crystalized for me one day when I was having a discussion with Heather. I don't recall what the exact issue was, but I was trying to explain to her, "No, Heather, we have to do it the way I'm telling you, because it's the law."

She replied, "The law is only a suggestion."

"No," I told her, "it's the *law*."

"No," she repeated slowly and deliberately, for emphasis, "it is only a suggestion."

Today I think back on those words with a chill. This cavalier attitude toward US law—which turned out to be prevalent at ZTE too—lies at the very heart of the cautionary tale that will unfold in these pages.

This attitude is especially concerning when you pair it with another central feature of Chinese business culture—its incredible sense of drive. Chinese employees work harder than anyone else on the planet. They follow the infamous 996 rule: you work 9 a.m. to 9 p.m. six days a week. And many consider that only the minimum. Chinese labor law stipulates a forty-hour workweek, but the law is essentially meaningless—or merely a "suggestion."

China is playing to win the long game, and losing is considered unacceptable. Its workers reflect this mentality. As an illustration, I recently saw a CNN Business report about Huawei. Its cover photo depicted a Chinese engineer lying watchfully awake on the rollaway sleeping cot beneath his desk, his open eyes lit by the glow of his keyboard. This photo perfectly encapsulates the Chinese work ethic, at least as I observed it. Another example is the campus of Chinese telecom Foxconn, where, according to *The Guardian*, netting is strung around the dormitories to catch employees jumping out of windows due to low pay and abusive hours.

On a social level, much was to be learned as well. For instance, Huawei's annual holiday party was held at a very run-down hotel ballroom and included light Chinese background music as the guests feasted on standard rubbery chicken banquet fare. But none of us were prepared for the "dessert" in store for us: a traditional tai chi dance performance (including a smoke machine intro) by none other than our Huawei USA President decked out in full Chinese garb. Can't make this stuff up. But hey, Donna and I did win the $100 Visa gift card raffle prize. And by the looks on my Chinese colleagues' faces, they were less than thrilled.

Another cultural element to assimilate was the Chinese mooncake. Each fall during the Mid-Autumn Festival, these calorie-bombs are offered to friends or at family gatherings. Typically the box they come in is gorgeous. But looks are deceiving. Inside is a dense pastry with a Chinese symbol embossed on it, filled with a thick, heavy layer of red bean or lotus seed paste. Yep, it's the Chinese equivalent of our American fruitcake: a holiday punchline rather than a delicacy to be consumed by actual humans. My Huawei office became a veritable mooncake redistribution center during the Mid-Autumn Festival.

But back to Chinese business culture, there was one other aspect of which I wasn't initially aware: the intense competitiveness between businesses within the same sector. Huawei and ZTE, for example, are at virtual war with one another. They love to sue each other and steal each other's customers. To have an employee taken away from you by your main competitor is seen as losing face. And the hugest feather in one's corporate cap, the ultimate ego-win, is when you can hire an executive away from the other company.

Could this really be happening?

I had barely started at Huawei when ZTE started courting me. I was flattered, of course, but I wanted to keep my focus on Huawei for the time being, so I politely rebuffed ZTE for a while. But about a year after I joined Huawei, I got a more serious call from the in-house recruiter at ZTE. He was an American gentleman who later became a friend. He told me that ZTE was planning to create a General Counsel position, for the first time ever, at its US office in Richardson, Texas. Would I be interested in applying?

Would a starving dog be interested in a seared eighteen-ounce ribeye right off the grill?

I was beyond ecstatic. This was it. This was what I had worked and hoped for. My dreams were finally coming true! And I was not going to let this golden opportunity pass without giving it everything I had. No other candidate was going to out-prepare me, out-answer me, or out-interview me for this job. I was *the* candidate. And I would be the new ZTE USA General Counsel. Period.

I remember my main job interview with ZTE. Armored in my Uniform, I sat in a small conference room with three or four Chinese ZTE executives, including ZTE USA's CEO, Xi Kai, and its COO, Dai "Len" Ling, along with the American HR Director.

Xi was a tall Chinese gentleman who spoke very good English. He always had a puzzled or squinty-eyed look on his face, as if he questioned every word coming out of your mouth. He was the face-flushing, voice-raising, fist-slamming type. He split his time between our Texas office and his home in San Diego. Extremely political in nature, Xi was known to be the "throw you under the bus to advance his own agenda" type. Probably how he became CEO of ZTE USA. I'd been warned about him.

Len, on the other hand, required a downward tilt of the head to look in the eye and was more rounded in shape. He dressed in finer clothes and was soft-spoken. I'm not sure if this was because his English wasn't as good as Xi's or because he was more intent on listening than speaking, but Len always appeared to be calculating and processing all at once.

Xi and Len were asking me prepared questions in English and staring at me with folded hands and polite smiles. I had no idea if they were understanding a word I was saying.

I knew the etiquette well and had been following it to a tee: arrive early, use formal titles when addressing people, don't initiate any handshakes, pay attention to business cards, and, of course, sell yourself without sounding boastful. Focus on the company, not yourself.

So when Xi, in his squint-eyed, quasi-puzzled gaze, asked me to summarize why ZTE should hire me, I was prepared to tell him, in great modesty, why I was a perfect fit for ZTE, a veritable unicorn. "I think my broad range of generalist training will allow the company to do more tasks in-house, which will be good for efficiency and cost. My wide knowledge of US law, although imperfect, will allow my team to foresee potential problems and help the company avoid making costly legal errors. My previous experience in a China-based telecom company means that you will not need to train me in the technological aspects of the industry, nor in the protocols and etiquette of Chinese business culture. And my wife and I are both from Dallas originally and are happily settled here. We do not have any children. Dallas is our home, so you won't need to worry about me wanting to relocate in two or three years. I would be most grateful for this opportunity and would regard it as the beginning of a long-term career relationship."

I noticed as I spoke that Xi seemed to lose the squint-eyed look. In fact, after I finished giving my answer, I remember Xi and Len looking at one another and nodding almost imperceptibly. I must have done well.

But regardless of how good my answers were, I knew my biggest selling point of all—my pièce de resistance—was one I didn't need to mention at all: they'd be stealing me away from Huawei.

CHAPTER 4

And We're Off!

The offer from ZTE USA came in the fall of 2011, just a week after my fortieth birthday. The idea that I would be a GC for the US affiliate of one of the largest telecom providers in the world was beyond exciting. Even the lack of competitive compensation didn't dampen my enthusiasm. I'll never forget Xi and Len telling me, "It is quite an accomplishment to be GC at such a young age." Their offer letter hammered home the same message, in case I had missed it in person: *you are young, you really want this job, and you'll take the low pay we are offering just to get that title.*

All of that was true, alas. The offer was for barely more than I was making as Assistant General Counsel at Huawei. But Donna and I figured the compensation would go up after I proved myself. Either way, this initial GC gig with ZTE USA would help me land my next GC job with another company, and eventually my salary ship would come in.

But in order for you to fully appreciate the magnitude of this opportunity for Donna and me, I need to wind the calendar back a bit . . .

A new challenge

A year earlier, in September 2010, Donna had decided to go out on her own and start her own law firm. It was something she'd always wanted to do, and the time felt right. It was, of course, a risky move, as such things always are. She had nothing in hand except a $10,000 check from her one and only client, and her office would need to be in our house until she could afford more professional digs.

I remember that drizzly afternoon like it was yesterday: driving to Donna's office and helping her load dozens of legal boxes filled with her client files and other belongings into our cars as she set out to embark on her new journey. After we dropped her stuff off back at our house, we went to buy her a new computer, printer, and other office necessities.

I treated us to a glass of champagne to celebrate her bold move. "Here's to your new future," I proclaimed. "You've spent ten years helping build someone else's law practice. Now it's time to build your own!"

But honestly, Donna's starting a business meant losing one of our two steady salaries. We both had sizable student loans to pay off, and my salary was barely enough to cover all our expenses, even if nothing went wrong. We knew we were in for a bit of a struggle, but we figured we'd find a way somehow. We were optimists, and we were following "the plan"—which hadn't steered us wrong thus far.

A few months after Donna's striking out on her own, early in 2011, we were making dinner one evening when her phone rang. Sometimes you can tell, just by hearing a ringtone, that a call is going to be an important one. This was one of those times.

Glancing at the caller ID, Donna looked concerned and stepped into living room to take the call. All I could hear was one side of the conversation, but it was all I needed to hear.

"Charla?" Donna said. Her sister in Colorado. "I've been waiting for your call. You did? So what did they say? *What?*" Suddenly, Donna began pacing frantically; I could hear her footsteps from the kitchen. I stepped away from prepping dinner and moved into the living room, leaning on the doorway, listening in. "Well, are they sure? Did you get a second opinion?" she said into the phone.

Second opinion? Uh-oh.

Donna listened for several seconds, her eyes rolling in her head as she paced back and forth. I then heard her say, "I love you, Charla. I am going to be with you every step of the way. It is going to be okay. We will get through this together. I'll call you back tomorrow. I love you."

She pushed the End Call button. Her face went white as she rushed to hug me.

I tossed the chopping knife back on the counter and reached for her. "What is it, Donna? What's happening with Charla?"

"Oh God, Ashley. She has breast cancer."

"Oh no. How bad? I mean . . . how long? I mean . . . what's the prognosis?"

"Bad, if not close to the worst possible."

Donna gripped me tighter, her face mashed into my shoulder, and she started to cry.

Over the next twenty-four hours, Donna pulled it together, as she always does, and went into warrior mode. She was determined not to sit on the sidelines and watch the situation passively. She wanted—needed—to do something to help. We both did.

At the time, Charla had three little boys, including a toddler, and now she was going to need to undergo a double mastectomy, extensive chemo, and radiation treatment. And so Donna and I decided the best support we could lend the family was for Donna to fly to Colorado for one week of every month to help take care of her sister, the house, and the three young kids.

We knew this was the right thing to do—the only thing to do, really—but, still, it represented a pretty significant challenge for us, practically and financially. Donna had just started a business and was trying to build a clientele. Now, by taking a week off every month, she'd be essentially taking a twenty-five percent cut in what little she was making, plus stalling whatever career momentum she was trying to build. And of course, we would need to pay for those flights to Colorado every month. I can't say it was the biggest financial challenge we ever faced—that was yet to come—but it was the toughest challenge we'd dealt with as a couple up to that point.

We started selling everything we could part with—on eBay—just to buy the plane tickets and supplement our wounded income. A lot of those things were luxury items we had acquired in our early lawyering days. When we got our first jobs in law, we both treated ourselves to some nice things. We didn't have the expense of kids, and we figured our careers were only going to improve as time went on, so we weren't particularly stingy with ourselves. If Donna wanted something—like a Dior dress or a Chanel handbag—she would often buy it for herself, just as I might treat myself to a Ralph Lauren Purple Label suit. We accumulated a fair amount of stuff in our early years, and now we found ourselves unloading it all.

Donna had grown up in the Dallas–Fort Worth area, like me. And also like me, she'd lost her father when she was young. The one thing we'd both gotten from our dads was the understanding that if you wanted something, you had to work for it and pay the price. Looking back now, I remember that acquiring those possessions had felt like a goal for us: a challenge that always had a "price" beyond the price tag. For Donna it meant putting in extra time. *Want those Dior shoes? That'll be an extra twenty hours on top of what you're already working. You want it, you gotta pay the price.* It made us appreciate owning those things more than if they had just been handed to us. We'd earned them, damn it. Now, though, under the present circumstances, those things no longer held the same value as before.

I remember one day going through the closets and drawers of our bedroom and making a call on every single item.

Chanel bag from two years ago? Sell.

Christian Louboutin pumps? Sell.

Prada sunglasses? Keep.

YSL bracelet? Sell.

Sterling silver cufflinks? Sell.

Edward Green shoes? Keep.

Mahogany jewelry box? Sell.

Antique shelf clock? Sell.

The Sell pile kept growing larger and larger as we thought about what Charla, our brother-in-law, John, and their three boys were going through. Our loss was only things.

We took photos of our stuff and posted away on eBay. Often, when Donna was away in Colorado, I would hold garage sales, where I would quietly unload additional household items in an effort not to call Donna's attention to the loss. She needed to keep her energy positive.

The financial challenge wasn't the hardest part. The emotional strain was worse. Donna would start to get extremely blue for several days before departing for Colorado on her monthly excursion. She'd usually be in full tears by the time she got to baggage check. Donna is a fighter, and it was killing her to know what her sister was going through and to be powerless to fight it.

"I hate this, Ashley," she would say. "I hate this so fucking much."

When Donna was in Colorado, she would walk Charla's older kids to school in the morning, then return to the house to clean and cook and help take care of the toddler. Her chores included picking up dogshit in the backyard; packing lunches, snacks, and backpacks; entertaining the toddler; making dinner; and helping with homework. And doing laundry. Tons of laundry. At first Donna told me how their washing machine had a cute little chime that let you know when a cycle was done. By her fourth trip to Colorado, she was saying, "If I hear that goddamn chime one more time, I am going to scream."

That was the easy stuff. Much harder were the quiet times alone with Charla, trying to offer her hope and encouragement while feeling powerless herself, and the seemingly endless rehab sessions in which she had to coax Charla through painful therapy routines she wasn't even sure had any real value. During that year, we would save what little money we could, and Donna would make trips to Neiman Marcus to buy Charla treats. These were little rewards for Charla, to coax her through her long hours of daily rehab. Donna would literally bribe her sister with them:

"Just two push-ups and the YSL lip gloss is yours." "Eat a strawberry and a slice of banana and you get the Tom Ford eye shadow."

Donna would always come back from these visits feeling drained and emotionally wrung-out, so I made a habit of bringing champagne on ice in the car with me when I went to pick her up at the airport. Maybe it was silly on my part, but it became a monthly tradition. It was a trying time in general, and we spent a good amount of it in real emotional turmoil.

Deliverance

With that as a backdrop, my offer from ZTE couldn't have come at a better time. And it couldn't have meant more to Donna and me—from a "destiny" perspective. ZTE wanted me as its General Counsel! Holy crap! In the midst of such a challenging period of our lives, this opportunity felt like our hopes and dreams coming true. It felt like deliverance, like a red carpet being rolled out to the life we'd been working toward since our first days at Loyola. All those student loans we'd taken out. All those jobs where I'd sacrificed money to gain experience. All those weekly bags of groceries Donna had brought me. Maybe now, finally, it was payoff time. Maybe I was finally going to reap after all those years of sowing.

So without any negotiation or further due diligence on my part, I signed my acceptance letter.

The fact that the offer coincided with my fortieth birthday added a whole extra layer of meaning to the equation. Your forties and fifties are the heart of your productive adulthood, your career maturity years, and this opportunity had arrived right on schedule. It seemed like a very grownup job, arriving just as I was embarking on my very grownup forties. The job looked and felt like a win. So perhaps we imbued it with more significance than we should have. We truly believed this was the dream job that would carry us into the future we'd both been envisioning for so long.

The day I got the ZTE offer, Donna went out and bought a cake (from my favorite bakery, A&J's, in my favorite flavor, Italian cream) with "Congratulations" across the top, along with a bottle of Veuve Clicquot... or two. We celebrated with abandon, our heads and hearts bursting with optimism. I remember Donna saying to me with her third glass in hand, "Ashley, you just turned forty, and you are now the General Counsel of a multibillion-dollar international telecom company. You have the world by the tail."

And I believed it. I had arrived at last. Things could only get better from here.

Starting at ZTE

The day the offer came, I began immersing myself in all things ZTE.

ZTE is a bit smaller than Huawei but is still a massive company. It is the second-largest publicly traded telecom company in China, behind Huawei, and the fourth largest in the world. The company was started in the mid '80s under the name Zhongxing Semiconductor Co., I learned, and changed its name in the '90s. In its early days, ZTE made semiconductors and modems and then wisely branched out into phones and, later, telecom equipment in general and mobile infrastructure. The company was known for spending a ton of money on R&D and owned more than 13,000 patents in the US and worldwide. If there was a mobile network anywhere on the planet, there was a good chance ZTE equipment and technology was embedded in it. The company employed upwards of 85,000 people worldwide and had annual revenues of around $713 billion.

And just as with Huawei, I was actually hired by the North American subsidiary, ZTE USA. Its headquarters were in Richardson, Texas, near Dallas, where it had its name on a large office building and occupied two floors. The Richardson site paled in comparison to ZTE's worldwide headquarters in Shenzhen, China, which boasted a city-sized campus complete with its own bus system, tennis courts, soccer fields, restaurants, and (today) high-rise hotel, but it was my local port of entry into the larger world of ZTE. And I couldn't have been more excited.

My first day of work was October 10, 2011. Decked out in my Uniform as always, I parked my car and walked under the entrance archway and up the stone steps. I felt like a hero of myth crossing the threshold into a magical realm. My future was unfurling in real time. I literally had to pinch myself. *Is this really happening to me? Seriously?*

A woman from the HR office greeted me at the reception desk. "Good morning, Mr. Attorney Yablon, and welcome to ZTE. Won't you please come with me?"

She handed me my new-employee packet and led me down a hall, where I was in for a few big surprises. In my fantasies of being a General Counsel, I had imagined I would have a couple of paralegals and a legal secretary, as well as a damn sweet office and the freedom and wherewithal to hire at least some of my own legal staff. Kind of like a head football coach who gets to pick his assistant coaches. That fantasy bubble quickly burst.

My office was modest at best—a generic office space like all the others that lined the outer four walls of the entire floor. It was a 9'x 9' room with a modular L-shaped desk and credenza. A desk chair. No other office chairs. Generic blank white walls. No art. No plant. Nothing else. An office for a mid-level claims adjuster at an insurance company.

But I had asked (remembering what I'd learned at Huawei) for a big whiteboard, along with tons of dry-erase markers of every color. And they had delivered—a huge whiteboard, 6'x 8'. That beast took up nearly a whole wall. These items would turn out to be more valuable than anything else I could have been given.

What I didn't know yet, but was soon to learn, was that, to the Chinese, "GC" was merely a title. Being an attorney was not as revered as it is here in the US. I was just another worker bee for the Queen Bee back at the Mothership in Shenzhen. On that day, though, looking out over the sea of cubicles covering the entire floor, I chose to feel lucky. *Hell, I'm in the US,* I told myself, *and in this country being a GC means something.*

But I was genuinely disappointed with the staffing situation. "Your staff is already in place and they are waiting to meet you," said my HR host. *Oh?*

She led me to the office across the hall where three young Chinese lawyers, all in their twenties and early thirties, were waiting for me, standing in a row. She introduced them to me by their full Chinese names, then said, "but you may call them Zhang, Matthew, and Meghan." Matthew and Meghan bowed and smiled politely. Zhang frowned and emitted something between a grunt and a dismissive sigh. I wasn't quite sure what the handshake protocol was in this situation, so I didn't initiate any hand contact, and neither did they. It was an awkward moment.

"Let's have a meeting in my office at ten o'clock," I told my new staff, "after I've had a chance to settle in."

The power hierarchy was such that I reported directly to Mr. Guong in Shenzhen. He was General Counsel for all of ZTE, worldwide, including all its subsidiaries. His second-in-command was also named Mr. Guong (I immediately dubbed them Guong One and Guong Two in my mind). I was connected by a dotted line to Xi, the CEO of ZTE USA and also, to a lesser extent, to the CEO of ZTE worldwide.

Of course, the Chinese executive team was on the opposite side of the globe, which meant direct contact with them would be limited due to time differences. So I didn't have a direct "boss" in the US offices, but CEO Xi came to my office on day one and did his best to welcome me in his fashion.

"As you know, Mr. Yablon," he said, in his well-spoken English and booming voice, "We have not had a General Counsel position before you, so this is a first for all of us. We know there will be a learning . . . arc for you in your new role. So please—not to worry. Take time. Be comfortable. Do not rush yourself."

His assurances set my mind at ease—the early days in a new job are always anxious ones.

I vividly recall my first official meeting with my new team. I knew the Chinese respected formality and protocol, so I decided to begin with a formal presentation of sorts. I remember standing there in front of my massive whiteboard, trying to explain our workflow and hierarchy while my class of three sat facing me like dutiful students, writing down everything I said but not asking any questions. The whole scene was exactly as comfortable as it sounds.

As the day wore on, I learned that my new team members had varying levels of English proficiency and work experience, but uniformly poor knowledge of US law. None of them were even licensed to practice law in the United States!

Why was this important? In the US, we use the term "unauthorized practice of law" when a person purports to provide legal advice without being either a licensed attorney or under the supervision of an attorney. Legal assistants and paralegals must have their work reviewed and approved by a licensed attorney. Similarly, my Chinese legal staff would be required to have all their work reviewed by me. Or maybe that step too was considered merely a "suggestion."

Had I been given the leeway to hire my own staff, I certainly would have wanted a Chinese national or two on my team—of course—but I would also have wanted some US-trained attorneys with good knowledge of US law and the legal credentials to practice in the country where our office was located! And maybe also a buddy or two to watch my back. It seemed once again I would be playing the role of teacher, explainer, and go-between.

Over the course of the first few days and weeks on the job, I found myself having to explain US law in very basic terms to my team members, just as I had at Huawei. I was shocked at how little they knew, not only about the letter of US law, but also about the spirit, the logic. They had that same lack of fundamental understanding, that same dogged sense of "law as suggestion" I'd encountered before. I was as baffled by this attitude as I'd been at Huawei.

My intrepid trio of legal eagles consisted of:

Zhang. Just Zhang. He didn't use an Americanized version of his name; he was above that. Zhang had been running the legal department (without a US law license and without attorney supervision) for quite some time before I arrived on the scene. He was used to making all the decisions and doing things his way. He didn't like the fact that I'd been appointed GC, and he didn't like me. Zhang was arrogant and hardheaded. I was later told he got the job at ZTE based only on his family connections. Once I learned that, his flippant attitude made more sense. He preferred to work without oversight, and he made it clear to the other two lawyers that he intended to continue doing what he wanted regardless of what I might say or do. Zhang spoke middling English. He was extremely short, wore glasses, chain smoked, and always reeked of cigarettes. His reputation as a womanizer was difficult to fathom, but it was one that persisted nonetheless. Zhang was married and had several children.

Matthew. I'm not sure where the "Matthew" came from, as opposed to, say, Chuck or Hugh, which would have made sense based on his Chinese name. Matthew was the youngest of the three. Fresh out of law school in China, he had bad English and equally poor skin. He was a pleasant enough fellow to work with and very agreeable, but he was also quite naïve and inexperienced in the world. Matthew was extremely tall. He and Zhang made a striking pair, visually speaking. Matthew took all his cues from Zhang and did whatever Zhang told him to do.

Meghan. Last, but certainly not least, was Meghan. Meghan was in her early thirties. She was short and fireplug-shaped, with dark, defining glasses and short-to-mid-length hair. She had a sweet and funny laugh that endeared her to me instantly. She also spoke the best English, by far, of my three young attorneys. Meghan was intelligent and generous, and, thank God, motivated to help me. She was keenly tuned in to the politics of ZTE as a whole and to the local office grapevine. But she was quite private about herself. It wasn't until I'd worked with her for many months that I learned she was married. She and her husband saw each other only once every four to six months, owing to their respective work situations.

One thing all three of my young Chinese lawyers had in common was that they put the 996 club to shame. What they lacked in legal training and US law credentials, they made up for in long hours and stamina. After working a full day for me, they would proceed to spend a major chunk of each night speaking on the phone to the legal team in Shenzhen—which, of course, was on the opposite side of the world and the clock. With what little time they had left in their day after work, they would hurry home to their small apartments near the office, which they shared with other ZTE employees, and catch a few Z's. And then the next day, they'd rinse and repeat. Not much of a life from an American point of view, but who was I to judge?

I can't stress this enough: Meghan was my lifeline at ZTE, and I honestly can't even imagine what my experience working there would have been like without her help. She was my go-to person when it came to navigating company protocols and procedures, such as how to submit an expense report or how to make sure outside attorneys got paid. She was also my translator, not only in a literal sense, but in a cultural sense too. She always took care to explain to me what was really going on during complex business interactions and behind the polite smiles.

For example, one night (at Donna's urging) I invited the team out to dinner, because, well, that's what you do when you're the new boss. I took them to a local restaurant where I announced, "I want to share with you some of the classic foods we Americans—well, at least we *Texans*—really enjoy. With your permission, I will order for the table."

"That sounds wonderful," said Meghan.

"Excellent! Great idea!" echoed Matthew. Zhang, of course, remained silent.

If you are from the Dallas area, you know all about Snuffer's—famous for their burgers, and the alleged creator of the original cheddar fries. I proceed to order burgers, Snuffer's famous cheddar-fries, grilled marinated chicken breast, and fried pickles. Obviously not something to ingest as part of a daily diet, but signature Dallas fare that I thought would give my guests a nice sampling of casual Texas dining. They smiled and laughed throughout the meal, praising the food and service.

"Delicious."

"Wonderful choice of restaurants."

"Thank you for sharing this experience with us."

Yay, I had scored a hit.

It was only later that Meghan revealed to me, in confidence, that they'd all become sick afterwards. I guess the 3,500-calorie cheddar fries were a bit too rich for their unaccustomed stomachs. Zhang, I learned from Meghan, had raged at her, "How dare he take us to such a terrible place and make us eat such garbage?" Zhang always had a great deal to say behind my back. Not so much to my face. Again, Chinese social culture was on vivid display.

In another early situation, pretty hilarious in retrospect, we were meeting with some ZTE executives about a piece of ongoing litigation involving ZTE and another corporation. One of the Chinese executives asked me, "Mr. Yablon, what is your frank opinion of this case? How do you honestly feel about our legal chances?"

I replied, "Well, all things considered, I'd much rather be in their shoes than ours."

After the meeting, I observed the same Chinese executive speaking to a couple of his colleagues in the hallway. He was visibly upset. What had

I said or done? Perhaps I'd been too frank in my assessment of the case? I pulled Meghan aside. "Do you know what's wrong with Mr. Huang? Did I say or do something to offend him?"

Meghan said, "Let me ask." She quietly approached the group, spoke to them briefly, bowed, and then returned to me. "He is upset because you insulted his shoes."

Oh jeez, my figure of speech about the shoes! From that point on, I learned to speak and write very literally, and to avoid using metaphors and idioms as much as possible. Pictures, drawings, and whiteboards (almighty whiteboards) became my go-to.

By far the most helpful thing Meghan did for me was to report on each night's conversations with China. As I said, every night my young legal team would spend hours in phone meetings with the GC's office in Shenzhen and with other executives from ZTE China. I was not invited to participate in these meetings and was given no official reporting of what went on in them. Strange, right? You would think the company's leadership would want their General Counsel to know everything possible. Information is power (but obviously, withholding it is even more powerful).

How could I be expected to protect the company legally if I had no idea what was going on inside it? The situation reminded me of the way some criminal clients lie to their defense attorneys about their crimes and are then dismayed when the truth come out in court and blindsides their lawyer. Your attorney is the one person you want to be completely honest with. ZTE's failure to do so was telling and became a major theme in this story.

It began to feel to me as if there was an actual orchestrated plan in place to keep me in the dark. Luckily, I did have Meghan. She would fill me in on what was going on in the company, both in China and in the US.

Chinese "magic"

A lot of what Meghan told me was confidential. I wasn't supposed to have the information, and I certainly wasn't supposed to be getting it from her. One day, not long after I started on the job, Meghan took me aside and said to me, "We need a code word I can use when I'm sharing confidential information with you."

"Okay," I said, "What word do you want to use?"

"Magic," she replied immediately. She'd clearly put some thought into this.

Okay, magic it was. Often when she and I were in the middle of a conversation, Meghan would casually say, "magic," and then drop a bomb like, "Guong Two thinks Zhao is an idiot and wants to see him replaced. He's going to leak an email next week that he thinks will get Zhao fired." Then we'd go right back to talking about humdrum contract details.

The power of this "magic" was huge. I received incredible intel from Meghan. I would have been absolutely in the dark without her.

But even Meghan's allegiance had its limits. I sometimes got the impression there were things she knew but would not or could not discuss with me. This was immensely frustrating. I was supposed to be the boss, but I had to constantly scramble to obtain even basic information on certain matters. Within my first few weeks at ZTE, I began to get a sense that something big and unpleasant was unfolding at the company, and that everyone knew more about it than I did. Something was about to go down; I could smell and taste it. That was the strange part, knowing something is about to happen. You don't know what it is or when it will go down, but you *know*. And the worst part? There's nothing you can do about it. So you just wait.

More than once, allusions were made to the timing and purpose of my hiring, which made me suspect I had been hired at this particular time for a very specific purpose—a purpose that wasn't being discussed with me. When I would ask anyone about this, they would say something like, "No, no, you were hired because we needed a great General Counsel and because we wanted to lure you away from Huawei!"

And then they would smile at me.

CHAPTER 5

The Thanksgiving Meeting

I spent the first few weeks on the job reviewing the company's pending litigation matters and trying to identify any other potential legal issues that might be afoot, while also familiarizing myself with the company power structure and the art of getting things done at ZTE. I'd only been at the company five or six weeks when "the Thanksgiving meeting" occurred.

The date was November 23, 2011, the Wednesday before Thanksgiving. I was told to report for a special gathering in the company's largest conference room—the one that opened off the lobby but that I'd never seen used for a meeting. It had a massive table, nice leather chairs, and floor-to-ceiling windows that looked out on the North Central Expressway. It was our "special occasion" meeting room.

I speculated that the date of the meeting might be significant. *Maybe it's a pre-holiday meeting to discuss end-of-year celebrations!* my brain optimistically proposed. Somehow I didn't think so. Not judging by the hushed hallway conversations and the dark, urgent looks in people's eyes. Something big was brewing.

The room began filling up with somber-looking executives. Hewing to proper Chinese protocol, I took my seat in the correct order. Ten or twelve of us were ranged about the table. In attendance were the CEO Xi, the COO Len, the CFO, several VPs, and a couple of other top executives. All of ZTE USA's top brass—including me.

Here is a little chart of the attendees:

Attendee	Title	Country of Origin
Xi Kai	ZTE USA/ CEO	Chinese citizen
Dai "Len" Ling	ZTE USA/ COO	Chinese citizen
A whole bunch of others (8 or 9)	ZTE USA/ C-level Executives	Chinese citizens
Ashley Yablon	ZTE USA – General Counsel	US citizen

Notice anything? Everyone was a Chinese citizen except me. Why did this matter? Because if things ever went south in a legal sense, these Chinese citizens could board a plane, fly back to China, and face no consequences or accountability in the US. They had no skin in the US game because our government has no jurisdiction over visiting Chinese nationals. But me? I was the only executive who was a US citizen. And who better to represent the company—if things did go south—than a US national attorney? Lily-white, young, US born and bred. Such thoughts passed through my mind as I looked around the table that morning, but I had no reason to give them any real credence at this point.

The attendees spoke varying levels of English, but mostly Chinese. Still, the meeting was conducted largely in English, either in deference to the fact that I was in attendance or because US meetings were held in English as a matter of course. The two people who took the lead were Xi (the CEO) and Len (the COO). Len's official title was Chief Operating Officer of ZTE USA, but Meghan had informed me that despite his relatively modest title, he was actually a major player in ZTE worldwide. So my radar went up whenever Len spoke.

Here's the gist of the meeting as I recall it:

"A potentially troubling matter has come to our attention," Xi said, getting the ball rolling. "The United States House of Representatives Permanent Selection Committee on Intelligence, known as HPSCI, is launching an official government investigation into ZTE as well as Huawei."

A round of concerned grumbles cycled around the table.

"As we all know," Xi continued, "the United States has long been . . . unhappy about our company's expansion into the US. They view us as a potential threat to US national security. They believe that our phones and our technology, when embedded in the US telecommunications system, contain the potential for spying on US citizens and the US government. They are also concerned that the ownership of our company is tied to the People's Republic of China, which they believe might enable China to take control of this embedded technology, should it desire to do so."

Why the concern

A quick bit of history here. For years, both Huawei and ZTE had been jockeying to establish a footprint on US soil, and the US had been rebuffing their efforts in various ways. (It still is.) ZTE, when I joined it, was already on its second or third attempt to build a US presence, having been squeezed out on its previous attempts. But now, in 2011, both companies had worked their way past some of the early roadblocks and were doing extremely well in the US (and globally). And the United States was—justifiably—concerned about this.

Why? Well, first of all, Chinese phones were known to have built-in "backdoors," which permitted the operators of the technology, if they so desired, to collect vast amounts of data on the phone's users—data that could be used to violate the privacy and security of not only US citizens, but also government institutions. Virtually everything you did on your phone could theoretically be tracked—emails, texts, calls, online searches and purchases, everything. All smartphones actually have this potential vulnerability, and there is ongoing disagreement as to whether Chinese phones are more dangerous in this regard than phones made in other countries. But still, Chinese-made phones could definitely be used in nefarious ways.

Phones weren't the only issue—both Huawei and ZTE made a wide range of equipment that was integral to mobile infrastructure. Could this equipment be trusted? That was the question. Thanks to the Russian election hacking scandal of 2016, we've all seen the kind of damage that can be done when foreign adversaries gain inside access to our sensitive cyber and telecom technologies. And once you've embedded compromised equipment and software in a system, it is difficult, if not impossible, to fully extract it and reverse any damage caused.

Bottom line: the United States government was being extremely vigilant about whatever foreign-made technology it permitted to become part of our telecom system.

And our government did not trust China, plain and simple.

The following remarks, taken directly from the report issued by the House committee after its investigation, summarize the US's concerns. I promise you I won't subject you to a lot of legalese or government-speak in this book, but this is an instance where the committee's actual words are worth reading. They clearly spell out the concerns that lie at the heart of this story, concerns that persist, on a global basis, even today.

> The United States' critical infrastructure, and in particular its telecommunications networks, depend on trust and reliability. Telecommunications networks are vulnerable to malicious and evolving intrusions or disruptive activities. A sufficient level of trust, therefore, with both the provider of the equipment and those performing managed services must exist at all times. A company providing such equipment, and particularly any company having access to or detailed knowledge of the infrastructures' architectural blueprints, must be trusted to comply with United States laws, policies, and standards. If it cannot be trusted, then the United States and others should question whether the company should operate within the networks of our critical infrastructure.

The risk posed to U.S. national-security and economic interests by cyber-threats is an undeniable priority. First, the country's reliance on telecommunications infrastructure includes more than consumers' use of computer systems. Rather, multiple critical infrastructure systems depend on information transmission through telecommunications systems. These modern critical infrastructures include electric power grids; banking and finance systems; natural gas, oil, and water systems; and rail and shipping channels; each of which depend on computerized control systems. Further, system interdependencies among these critical infrastructures greatly increase the risk that failure in one system will cause failures or disruptions in multiple critical infrastructure systems. Therefore, a disruption in telecommunication networks can have devastating effects on all aspects of modern American living, causing shortages and stoppages that ripple throughout society.

Second, the security vulnerabilities that come along with this dependence are quite broad, and range from insider threats to cyber espionage and attacks from sophisticated nation states. In fact, there is a growing recognition of vulnerabilities resulting from foreign-sourced telecommunications supply chains used for U.S. national-security applications. The FBI, for example, has assessed with high confidence that threats to the supply chain from both nation-states and criminal elements constitute a high cyber threat. Similarly, the National Counterintelligence Executive assessed that "foreign attempts to collect U.S. technological and economic information will continue at a high level and will represent a growing and persistent threat to US economic security."

Third, the U.S. government must pay particular attention to products produced by companies with ties to regimes that present the highest and most advanced espionage threats to the U.S., such as China. Recent cyber-attacks often emanate from China, and even though precise attribution is a perennial challenge, the volume, scale, and sophistication often indicate state involvement. As the U.S.-China Commission explained in its unclassified report on China's capabilities to conduct cyber warfare and computer network exploitation (CNE), actors in China seeking sensitive economic and national security information through malicious cyber operations often face little chance of being detected by their targets.

Finally, complicating this problem is the fact that Chinese telecommunications firms, such as Huawei and ZTE, are rapidly becoming dominant global players in the telecommunications market. In another industry, this development might not be particularly concerning. When those companies seek to control the market for sensitive equipment and infrastructure that could be used for spying and other malicious purposes, the lack of market diversity becomes a national concern for the United States and other countries. Of note, the United States is not the only country focusing on these concerns. Australia expressed similar concerns when it chose to ban Huawei from its national broadband infrastructure project. Great Britain has attempted to address the concerns by instituting an evaluation regime that limits Huawei's access to the infrastructure and evaluates any Huawei equipment and software before they enter the infrastructure.

So that was how our government was feeling about China at the time.

Back to the Thanksgiving meeting...

Xi and Len explained that the US was focusing its investigation only on Huawei and ZTE—while ignoring other Chinese and foreign companies—for two reasons:

- First, because ZTE and Huawei were the biggest players in Chinese telecom manufacturing.
- Second, because Huawei itself had invited the investigation.

Earlier in the year, in response to its frustration about being continually hobbled in the marketplace by US suspicions and restrictions, Huawei had issued an open letter to the US, saying essentially, *Hey, come investigate us. We welcome it. Open our books. Visit our factories. We want you to see how clean and innocent we are. Come on, hit us with your best shot.*

Apparently, the US had taken them at their word. And ZTE, because it was so like Huawei in its scope, model, and products, was pulled into the investigation by association.

"We consider the targeting of ZTE to be unfair and unwarranted," said Xi with his thunderous voice, "but there is little we can do now but cooperate with the investigation. As you can imagine, the outcome of this investigation will have massive consequences on both ZTE global and ZTE USA."

Massive indeed. Think about it. If the US government were to conclude that ZTE was not a trustworthy technology partner, it might very well ban the sale and use of ZTE products in the US. Why would that matter so much? Because even though ZTE and Huawei sell their phones throughout the world, selling their phones in the US is the ultimate coup. It's the reason everyone wants to be big in the US: prestige. And money too. Lots and lots of money.

So how do ZTE and Huawei manufacture their phones? They buy the best technology they can find for each part of the phone. Not surprisingly, the best technology is often made by US manufacturers. Want the fastest processor for your phone? Buy the US processor. Want the best screen technology? Buy the US screen. You get the idea.

Want to know why the Chinese are better than anyone else in the telecom market? It's a simple equation: inexpensive Chinese manpower/

manufacturing + the best US technology = least expensive competitive product. The Chinese have cheap manpower—cheaper than the US and most any other country. Couple that with the best US component parts and you rule the marketplace. And China does.

Now imagine if ZTE or Huawei phones were not allowed to be sold in the US. That could represent the loss of hundreds of billions of dollars to ZTE as a whole. Or imagine if the US telecom manufacturers (makers of processors, screens, et cetera) were forbidden from selling their component parts to ZTE or Huawei? Think of the money that would be lost in the US. Billions again. Billions of dollars in trade. Hundreds of thousands of US and Chinese jobs gone.

And now consider the fate of ZTE USA, apart from ZTE as a whole. You see, ZTE USA wasn't involved in manufacturing at all. Its sole *raison d'etre* was to sell Chinese-made phones in the US, largely by contracting with the big carriers like Verizon, T-Mobile, and Sprint. ZTE USA procured the deals and ZTE China shipped over "white-labeled" devices, which allowed the US carriers to put their own names on the products. But if ZTE's phones were no longer allowed to be sold in the US, there would be no reason for the continued existence of ZTE USA. The company would have to fold and retreat. Once again. So an unexpected investigation by US authorities was, needless to say, a very big deal. On many levels.

"The question is, what do we do about it?" asked one of the executives.

All eyes landed on me.

The Washington angle

Luckily, I'd had time to prepare my answer. "Okay, here's what we need to do," I said. "We go directly to Washington and hire the best lobbying law firm money can buy."

The executives stared at me in frozen perplexity. Then, as if on cue, they all stood up from their chairs, walked away from the table, and huddled in the corner of the room, leaving me alone at the table. Awkward? Nah.

Over the next thirty seconds or so, they hastily exchanged words in Chinese. One of them turned to me at last and said, "We do not understand. We hired you to be our General Counsel. Why can you not handle this matter for us?"

Affecting more calmness than I was feeling, I tried to explain to them that I was ZTE's *General Counsel*—not an expert in every type of law. I then gave my standard spiel about how US attorneys are specialized, and you don't hire a divorce attorney to sue a giant chemical corporation, blah, blah.

"This case has massive national and international implications," I explained. "It is political in nature. We need Washington access, and we need high-level PR and strategy. We need lobbyists."

Once again, there was a brief huddle in Chinese, then Xi, with his typical puzzled expression, said, "Please tell us, Mr. Yablon, what is this 'lobby' you go on about?"

This was going to be a tough one. I yearned for my whiteboard.

I gave them a two-minute crash course on national politics, explaining that political influence and access are, alas, essential to moving the needle in Washington, and that lobbyists are the ones who know how to play that game. I also explained that lobbying is a high-level competency that most law firms don't possess. Only the elite Washington firms have lobbying capability.

The executives didn't seem to fully understand. They went into another powwow, speaking in Chinese once again. Confusion reigned.

Finally, one of the men's eyes lit up at something being said and he faced me with a smile. I was starting to think I had made a dent in his understanding until he said, in halting English, "We watch your C-SPAN. Person stands at microphone and testifies on The Hill. We want that person to be you. You testify on Hill." The others grunted and nodded enthusiastically. Uh-oh.

A giant red flag had just been waved in my face, but it was so blatant I didn't even see it at the time. All I remember thinking was, *Perfect. These guys haven't heard a word I've said, and they have no idea how our government or legal system work. But somehow, they're all experts on C-SPAN?*

I could tell from the expectant faces of my bosses, though, that they had put some thought into this absurd notion they were proposing. They seriously imagined I was going to travel to DC armed with documents and charts, stand up in front of Congress, swear under oath to tell the truth (well, their "truth"), and deliver a happy spiel from the podium about the virtues of ZTE—and that somehow my testimony would magically fix the whole problem and make it all go away.

Again, all eyes were on me. I could feel myself sweating through my Uniform. I explained once again, with greater emphasis (and less patience), why such a simplistic notion wouldn't work. Eventually I was able to get through to some of them, and they came to grudging agreement with me: we needed to take this matter extremely seriously, and we needed to hire the biggest, baddest, most politically connected DC lobbying law firm we could find.

CHAPTER 6

More Red Flags

Be careful what you wish for. I was tasked with the job of vetting all the major lobbying law firms in Washington and then presenting a list of my top five recommendations to the ZTE brass. And so I spent the rest of November and all of December and early January vetting law firms—while also managing all of the company's litigations and other legal business.

Now, vetting law firms might sound like a pretty non-taxing chore, but it was actually a ton of work. When hiring a big law firm, you don't just look at their website, read a few testimonials, then call them up and ask whoever answers the phone, "Hey, how much do you charge to represent a global telecom company being investigated by the HPSCI?" First you have to find the right partner to talk to, then you need to describe your legal situation in detail and answer the barrage of questions they inevitably have, and then you need to schedule further phone meetings in which they bring in their partners and other in-house experts to ask additional high-level questions. Pages and pages of questions. Then it's your turn to start asking them questions. Further phone calls ensue, et cetera.

Eventually, long-winded proposals are provided to you, and you spend hours reading through each law firm's legalese-laden caveats, multiple carve-outs, and various exceptions—clauses that basically amount to *Your payment to us is not a guarantee of our doing anything, and we make zero promises to you;* or *This advice is based only on what you are telling us and we cannot be responsible for blah, blah;* or *Under no circumstance will we, the law firm, in any way be liable to ZTE USA for anything under the sun* and so forth.

It's like the fine print at the bottom of a set of contest rules that no one ever reads (except attorneys)—the stuff that protects the offeror of the prize and basically says you've won nothing but have actually sold away all your rights. The very reason everyone hates attorneys.

After cutting through all that red tape, my job was to try to get at the real answers my ZTE brethren were interested in: what is each law firm's plan to get us out of this mess, and how much is it going to cost? Simple in theory. Virtually impossible to nail down in reality.

During this period, I found myself on the phone at all hours of the night and all weekend long, every weekend. The firms I was talking to were the absolute cream of the legal crop, the best-of-the-best, "silk stockings" firms. These were firms where the partners earned a thousand dollars an hour or more and the associates earned more than the partners do at most other firms. And then there were the swarms of highly paid paralegals, assistants, and legal secretaries you needed to pay as well.

Most of the firms I spoke to were willing to discuss their basic strategies and ideas with me. All except one firm—the biggest, baddest, and priciest one of all: DD&M. The attitude I got from them felt like, "We're DD&M; we don't answer questions, YOU do." I spent countless hours answering their queries, while receiving virtually nothing useful from them in return. Needless to say, they were not my favorite choice.

By the later part of December, Xi was putting pressure on me—literally yelling at me at times—to hire a firm. He was on the hot seat as CEO of ZTE USA, and he knew the HPSCI people would be breathing down his neck soon. He was starting to feel the pressure. I assured him that I would have my list of firms chosen soon and that we should have a team in place in January.

Meanwhile, on the home front, Donna was starting to build momentum with her private practice while still traveling to Colorado once a month to help take care of her sister. The latter situation continued to be sad and stressful, but things were finally looking up; Charla was responding positively to treatment.

We had been dreading Christmas of 2011, considering Charla's family circumstances. But now there was hope in that household, and so Donna was spending a lot of her time and energy trying to help her sister's family have a fun and joyful holiday.

Donna and I felt incredibly fortunate that Christmas. I mean here we were, one of us building a promising new law firm, the other having just embarked on a new corporate law career where the sky was the limit.

I remember that Christmas Eve, sitting with Donna in front of our Christmas tree and having a celebratory drink. It had been a damn rough year, but now, when we looked at our tree, we saw only optimism for the future.

We made multiple toasts that night, but one in particular stood out. I remember lifting a glass and saying, "To a new year. This past year has been tough on us all. Thank God Charla is better. And hey, here's to ZTE. I believe getting this job was a turning point for us, and I am so grateful." And like Bob Cratchit toasting Mr. Scrooge, I lifted my glass to my new employer.

Donna cuddled against me and clinked her glass on mine, and for the first time in a long while we felt nothing but hope. We could feel that 2012 was going to be a good year.

Another red flag missed

Another event happened in December that seemed only peculiar at the time, but that in retrospect should have been another red flag for me. One day Meghan approached me and said, "Ashley, will you talk to our import/export lawyer in Shenzhen? He has a question for you about US export laws."

"Sure," I replied, thinking nothing of it.

Meghan set up the call. The Chinese gentleman she wanted me to speak with went by the Americanized name Preston. We connected by phone and, after exchanging pleasantries, he said to me very matter-of-factly, in a soft and fairly refined English, "We would like to know if there is some way that ZTE USA can legally export products to countries such as Iran, Cuba, and North Korea, where there are US embargoes in place."

"Excuse me?" I said, nearly choking on my coffee. I assumed I had misheard him.

He repeated the question. No, I'd heard him right. He wanted to know if there was a legal way for ZTE USA to sell products to countries with which the United States had trade embargoes. Kind of like asking if there is a legal way to rob a liquor store.

"I'm no expert," I told Preston, as politely as I could, "but I doubt there is a way to do that. The whole purpose of an embargo is to make that kind of trade impossible. But, as I said, I'm not an expert. I'll tell you what I'll do: I'll find someone who knows more about this than I do, and we'll schedule another call."

"Very well, yes. Thank you."

I called a friend of mine, Rob, who was a partner in a large law firm, and he told me his firm had a department that dealt in import/export law. He put me in touch with their top specialist, a woman named Tracy, and we set up a four-way call between me, Meghan, Preston, and Tracy.

Preston essentially repeated the same query to Tracy: How can ZTE USA legally sell products to banned countries? Tracy asked him a series of clarifying questions, more to ascertain whether she was hearing him correctly than to tease out additional information or to lead him to believe this was even a possibility.

After the phone meeting was over, Tracy immediately called me back. "Let me get this straight," she said. "He's actually asking if there is a legal way for a US-based corporation to do something that is blatantly illegal under US federal law?"

"That's what I'm hearing," I owned.

"Before I tell him he's out of his freaking mind, let me gather some additional information from each of you. See if I'm missing anything. Give me everyone's email addresses, please."

Within the next hour, Tracy had crafted an email, containing a list of additional questions about import/export matters, and sent it out to me, Meghan, and Preston.

Almost instantly, a reply came back from Preston saying something to the tune of, "Never mind, we decided not to pursue that avenue. Please forget about it."

O-kay.

I sat back in my chair, contemplating the whole weird exchange. I was confounded as to the purpose and logic of Preston's initial query. *Why is he even toying with the idea of bringing ZTE USA in on a plan to sell to banned countries?* I wondered. *That's not what ZTE USA does. ZTE USA sells ZTE-China-made products to US companies like T-Mobile and Sprint; its focus is strictly US-based. ZTE China is the company's manufacturing center, not ZTE USA. If China wants to sell phones or other equipment to Iran or Sudan, why not do so directly? Leave the US branch out of it?*

The idea made no sense. So why the call? Was Preston on some fishing mission to try to better understand US trade law? Was he hoping to get this green US General Counsel (me) to suggest a new legal angle he hadn't thought of for getting around embargo restrictions? Was he simply testing me to see what my risk tolerance was? My ethical boundaries?

With so many other issues to attend to, however, I quickly moved on to put out the next fire.

In retrospect, I can only think that my (and Tracy's) quick negative reaction to Preston's queries told him exactly what he needed to know: that this young General Counsel, while green, was starting to catch wind of what they were hoping I would do for the company—and I was not willing to be their huckleberry. Hence the speedy "forget we said anything" response.

Off to China

In early January I took my first trip to the Shenzhen offices of ZTE.

Further red flags were waved at me on this trip. But before any of that happened, I was introduced to the joys of ZTE's employee travel policy.

ZTE behaved like a world-class organization when it came to the salary and benefits packages it offered its executives (myself excluded), but when it came to travel, it behaved like a Mom-and-Pop convenience store chain.

Everyone in the company was required to follow the same strict policies, and these policies appeared to have been written in 1957. All employees, regardless of position, had to fly coach. Okay, that was understandable and fine with me (though, as I was soon to learn, those coach seats turn into a unique form of torture device on a twenty-hour nonstop flight). The part that boggled the mind was this: ZTE granted employees a $25 per day travel allowance, which included lodging, meals, and local travel (taxicabs, metro, et cetera). Let that sink in for a minute. Can you imagine even trying to book a parking space in a major city for $25 a day or less? You'd be lucky to rent a refrigerator box in a homeless camp for that amount. Seriously. And then you were supposed to have money left over to cover meals and transportation. It defies belief that this was the actual travel policy of a twenty-first–century corporation vying for world-class status. Based on my meager salary, I figured the travel stipend would be modest, but this was ridiculous.

Dallas to Hong Kong takes anywhere from seventeen to twenty-five hours, depending on whether your flight has a stop or not. Hong Kong is the closest and cheapest port from which to cross over into China, and Shenzhen is the first city major city you encounter in China, once over the border.

After my nearly twenty-hour flight, I extracted my spine from the cramped seat, feeling like I needed an emergency chiropractic intervention. I proceeded to fight my way out of the airport, through the crowds, and spend half my per diem budget on a bus from Hong Kong, which then crossed the border into China and into Shenzhen.

My mood lifted as the bus neared Shenzhen. I was blown away by what I saw. Shenzhen is an astounding place. It's the fourth largest city in China, after Shanghai, Beijing, and Tianjin. But even at number four, its population exceeds New York City's by about fifty percent—not counting all the temporary and transient residents and visitors, which put Shenzhen's real population at around twenty million on any given day. The city is huge, bustling, kinetic. Shenzhen is the high-tech capital of China, the new Silicon Valley. Whatever crazy ideas the global high-tech industry dreams up, they probably make the hardware for it in Shenzhen. The astonishing thing about the city is how rapid its growth has been. It swelled from a town of 60,000 to a megalopolis of more than twelve million in less than four decades! Insane.

A big part of that growth has been fueled by the telecommunications industry. The two big telecom companies—Huawei and ZTE—have their global headquarters there, as do other high-tech giants like BGI, the genome-mapping company; BYD, the IT and automobile corporation; and Tencent, the multinational high-tech holding conglomerate.

The place is incredibly dynamic and full of energy. The people are young, ambitious, and tapped into the latest technologies. You can buy a melon from a street vendor by scanning it with your iPhone. Skyscrapers seem to pop up around you in real-time. Traffic flows like a rushing river. Each little district is a thriving universe unto itself—the financial district, the ex-pat zone with Western-style restaurants and entertainment venues, the bustling mega-village where the locals live and shop, and on and on.

My eyes were the size of coasters as I gazed out at this marvel of a city from my seat on the bus. Now *this* was what I had dreamt about being part of. When you're in Shenzhen on business, you can't help but feel you're part of the pulse-beat of something important and vital.

Which was exactly how I felt—until my bus pulled over. I was about to discover how it's possible to book a room in a world-class city like

Shenzhen and still have plenty of change left over from a twenty. You do so by staying in a place like Meet-in-the-Blue-Sky. Look it up. I'll spare the name of the actual place I stayed with its attached brothel and karaoke bar, but you get the idea. Let's just use Meet-in-the-Blue-Sky for our purposes. The closest American thing I can compare it to is a youth hostel, but even youth hostels in America are more luxurious than this place was. From the outside it was a run-down, nondescript fifteen-story building attached to a multi-story karaoke club. But my jaw almost hit the floor when I opened the door to my room. It made a cell in a Zen monastery look like the Ritz-Carlton.

There was no air conditioning, no fridge, no kitchenette, no comfy chair. The furnishings consisted of a hard office chair in front of a wall-shelf that served as a desk, a 1970s tube television, a toilet, a shower, a sink, and a "bed" that wasn't really a bed but a rattan-like mat on a metal frame, with no mattress and one thin sheet. The air was stuffy and had a weird, dingy smell. The walls were rice-paper thin. But the detail I remember most was the frosted glass partition separating the living room/bedroom from the toilet/shower. It was covered with little Winnie-the-Pooh stickers that some visiting child had apparently plastered there out of boredom and no one had seen fit to remove. An exquisite finishing touch. I had to take a picture of it to show Donna.

Interestingly, I never saw any other traveling ZTE employees while staying here, including Matthew, Zhang, or Meghan. Never. During any of my visits. And of course, they never told me where they were staying.

After the twenty-hour flight in coach, my first night on the San Quentin-inspired bedding was exactly the sensual delight you might imagine. I'd say I awoke sore and miserable, but that would imply that I actually slept. As I sat up on the edge of the "bed" and surveyed my environs in the unforgiving light of morning, the accommodations made me feel even more depressed than they had the day before. Why would a major player in the global telecom market choose to treat its traveling employees this way? Anyone who slept in a room like this was bound to wake up tired, sore, angry, disheartened, and resentful. Is that how you want your visiting executives to start their workdays? I was ZTE USA's General Counsel, for crying out loud, and they had me staying in a room an Appalachian Trail hiker would turn down.

So even though I was excited to be in China, I wasn't in the most chipper of moods. After showering and putting on my Uniform, I was already sweating profusely owing to the lack of air conditioning and the heat of the day. Breakfast would have to wait—especially on such a limited budget.

I needed to make my way to the ZTE campus. Meghan had emailed me basic directions: Walk several blocks down the street to the underground metro system, take the red line several stops to the ZTE campus stop, go back above ground, and walk several more blocks to the campus—to the large white main building. Seemed easy enough.

I made my way down the street to the metro station and went underground. So far, so good. But once in the station, how was I to find the train? Especially when all instructions were in Chinese? By watching a few travelers, I learned I needed a metro coin. After inserting some Chinese currency into the machine and pushing a bunch of buttons, a metro coin was miraculously spit out. I popped it into the turnstile and followed the crowds to the red line.

This is where it got odd. At rush hour, there are police stationed at the train doors whose sole job is to push and shove passengers aboard the train. Chinese travelers eagerly assist in this process as well. As you are forcefully shoved into the car from behind, you must brace to maintain your balance.

Once aboard the train, crammed in like a sardine and gripping the handle rail, I spotted a map for the red line, including all its stops. I noted that the ZTE campus exit was seven stops away. *Okay, hold on tight and prepare for exit*, I thought.

As the train braked at my eventual stop, my upstream exit against the downstream rush of onboarding passengers was nearly enough to rip the briefcase from my hand. I made my way to the stairwell, ascended the steps, and walked the last few blocks to the campus.

I must admit that seeing the campus did a lot to improve my spirits. It was indeed a city unto itself. Clearly this was a company that was absolutely thriving, and my future here had just begun. I decided I could deal with the funky travel accommodations if that was the price I had to pay for being part of this company's future.

The main white building was a literal ivory tower. This was where all the executives had their offices and where ZTE showed off its best accommodations to its visitors and customers.

As I opened the door to the ivory tower, Meghan was waiting for me in the lobby. "Have any issues getting here?" she asked.

"None," I lied. I am a team player after all.

We made our way upstairs to where the entire ZTE China legal department was located. It took up almost the entire floor and reminded me of a 1950s elementary school—poor fluorescent lighting, stale air, and a linoleum-looking floor. All that was missing was a beaten down chalkboard. That had been replaced by—you guessed it—whiteboards!

I was assigned a cubicle to work in, not a guest office, which again made me feel like a temp accountant rather than the company's main US attorney. Shortly after logging in on my computer, I decided to stretch my legs and explore this ivory tower a bit.

In the course of my unguided walk-through, I learned that the floor beneath the legal department was known as the Hospitality floor—the floor with really comfy air conditioning, a welcome area with a high-end coffee and espresso machine, and a cadre of young, good-looking male and female ZTE Ambassadors. This was where visitors and customers were taken. During my stay, I made it a point to visit the Hospitality floor often. But despite the floor's niceties, the restrooms took some getting used to—not just on the Hospitality floor but on all the floors, including the legal floor. There were urinals, yes, but no toilets. Oh, there were stalls with locking doors but when you entered one you saw only a toilet paper holder and a gaping hole in the floor. And the smell was just what you would imagine.

The visit went strangely. The official purpose of my trip was to discuss my performance since joining the company and to talk about my goals and projects for the coming year. But every meeting I had went something like this:

> **Me:** "So, let's talk about my role with the company as the new year unfolds."
>
> **Them:** "Have you been able to learn anything more about the HPSCI investigation?"
>
> **Me:** "No, but I'm monitoring the situation and I'll let you know as soon as I learn anything. Anyway, as I see it, there are three major initiatives for 2012 that I would like to—"
>
> **Them:** "Have you been contacted by any US government agencies?"
>
> **Me:** "No, but I have calls in to several of them. So getting back to my idea about—"
>
> **Them:** "How is the search for a law firm going?"

Me: "Great. I'll be sending you my recommendations shortly. But for now let's—"

Them: "Is there anyone you can call, today, to get more information about the investigation?"

My first night of work, just to assuage my hosts' anxiety, I made some phone calls to the US late in the evening (daytime in the US). I was able to get through to the GAO (Government Accountability Office) and managed to speak to someone who had some actual knowledge about the investigation. She told me the agency would soon be contacting several people at ZTE to ask questions regarding cybersecurity and licensing issues, at the bequest of HPSCI.

The next morning, I shared this information with my Chinese hosts. They seemed relieved to have some new intel to work with, but they were still weirdly apprehensive. Reading the looks on their faces, I again sensed I was not operating from the same set of facts as everyone else. There was something I wasn't being told.

On day two, I was finally due to have my first face-to-face meeting with Mr. Guong One, my boss and the General Counsel of ZTE worldwide. I had never met with him before, even during my interview process. I was excited and nervous at the same time.

Hewing to proper Chinese etiquette, I was fully (if not overly) prepared with paperwork documenting the goings-on back home at ZTE USA. With briefcase in hand, I made my way to his office and knocked on the door. Entering, I met a tall, thin, avuncular gentleman with glasses and a welcoming smile. Despite the warm weather, he wore a gray sweater vest. We exchanged a semi-bow—not shaking hands, of course. As I took a seat in one of the two chairs in front of his desk, he slowly walked back behind his desk, continuing to sip his tea. Mr. Guong spoke with a soft, almost whispering voice and impeccable English. He had the look of a philosophy professor, not that of a General Counsel in charge of a multi-billion-dollar telecom company.

Despite my preparedness for our meeting, I was asked only one question—but in twenty-five different ways: "How is that search for the lobbying law firm going?"

My responses were always a variation of the same: "I am making progress and will have my shortlist of five firms completed within the next couple weeks."

And then he would ask the question again. It was like the movie *Groundhog Day*; we were stuck in an endless repetitive question/answer loop. Not exactly how I envisioned my first meeting with The Great and Powerful Oz.

Wrapping up

I spent the rest of my trip trying to arrange meetings related to the alleged purpose of my visit—i.e., to receive guidance and feedback about my job from the head office—and being diverted instead into conversations about the investigation.

Whenever I would ask Guong One, Guong Two, or any other members of the law team whether ZTE had any legal or technological exposure pertaining to the investigation that I should be concerned about, they would all smile and tell me that, no, ZTE was clean in all areas, so there were no worries. Then they would go back to grilling me about the investigation.

Each evening I would return to Meet-in-the-Blue-Sky, where my only companionship was those creepy Winnie-the-Pooh stickers. I did go out for a few meals with colleagues, including Meghan. That was nice. Hey, she treated—but how was she able to have a company credit card when I was not?

After a few meals in Shenzhen, I quickly established a couple of favorites: (1) duck prepared any way, and (2) a great dish with a horrible name, Saliva Chicken. No joke, that is the name. It is Szechuan-style chicken poached in a chili sauce. Pun intended or not, the menu always uses the same description: "mouth-watering." I always chuckled.

But more often than not I found myself dining at the local KFC. Why? Because it was right near the hostel, and it was a venue I could comfortably navigate. I never eat fast food back home, but KFC was all my travel budget could afford.

Seriously, if you were to see me schlepping back and forth from the KFC to the youth hostel, you would no doubt conclude I was a traveling American novelty salesman down on his career luck, not the US GC of a multinational corporation. One thing I did discover, though, was that the chicken at KFC China tasted as fresh as it did in the US South, where the Colonel was king. Go figure. At least I was learning something useful on this trip. Five days in Shenzhen and that was about all I had to show for it.

As I settled into my coach seat, bracing myself for the day-long journey back to Dallas, my mind kept circling back to one red-flag question: *If everyone is so confident that ZTE is clean, why are they all so worried about this investigation?*

CHAPTER 7

It Hits the Fan

When I returned to Dallas from China, I continued to receive pressure from Xi, the US CEO, to get a law firm hired. During his frequent rants on the topic, I often witnessed his typical puzzled look replaced by the flushed, about-to-pound-the-desk look I've mentioned. And so, early in 2012—a few weeks after I got back from China—I submitted my top five recommended firms to the ZTE executive team. It had been a hell of a lot of work.

The Chinese team and I had some phone discussions back and forth about my shortlist. I made it known to them that DD&M was probably my least favorite choice, in spite of its global reputation. I'd put the firm on the list only because of its lobbying strengths, but the DD&M people had given me nothing but entitled attitudes. Some of the other firms had offered helpful strategies I thought we could work with, but not DD&M.

Xi, who really didn't know anything about US law and seemed primarily concerned about his own exposure, was leaning heavily toward hiring DD&M, strictly because of their name. He thought they would have the most clout. I'm sure he was doing his best to persuade ZTE China of his point of view, because every time I would try to offer my recommendations to Guong One, Guong Two, or others in China, I would be given the equivalent of a parent patting a child on his head and thanking him for his input. This was demoralizing, to put it mildly. I mean, here was this group of Chinese executives, completely removed from US politics and knowing nothing about US law, overruling my thinking in an area they knew next-to-nothing about. I, on the other hand, had lived and worked in the US my entire life, had served in a congressional office while in law school, and had been hand-picked as ZTE USA's own damn General Counsel—so why didn't they want the benefit of the expertise they had hired me to provide?

That nagging question again. I chalked it up to the mystery of their business culture. Chinese magic.

As the next several weeks unfolded, Xi grew more and more anxious about getting a law firm in place. ZTE China, for its part, was inexplicably dragging its feet. So I was getting pressure from Xi to make the hire, and all I could do was pester the Chinese executive team about it. I sent several emails to Guong One and Guong Two, pleading for them to move forward; my emails were essentially ignored. This went on for a couple of months.

And then suddenly everything changed.

The Reuters bomb

On March 21, I received a call from a person in the ZTE Public Relations office. "Are you sitting down?" the young man asked.

"Do I need to be?" I said.

"I would be if I were you. Reuters is going to drop a bomb tomorrow. It has to do with ZTE illegally selling US parts to Iran in defiance of US embargoes."

"What?"

"Stay tuned, gotta go."

The next day, the bomb indeed dropped. Reuters published a piece entitled, "Special Report: Chinese Firm Helps Iran Spy on Citizens." And it was a doozy. The article explained how ZTE, in a $130 million deal, had sold Iran a countrywide telecommunications system that was capable of spying on all Iranian citizens—by intercepting their voice calls, text messages, emails, web activity, and more. But the worst part of the article was its claim that many of the components of the system had been made in the

United States. Iran was under a strict US trade embargo, so if ZTE China was: (1) making telecom equipment containing US component-parts; and (2) selling that equipment to the banned country of Iran, they were doing something massively illegal under US law.

The allegation was a monumental one, which, if true, would not only cause a major shakeup within ZTE worldwide but would also create ripples across the whole telecom industry and US–China relations. This could be a game-changer of the highest order.

The really damning part of the article was this: Reuters had also gotten hold of the actual packing list of the shipped equipment—all 907 pages of it.

You know what a packing list is, right? It's the type of thing you get when you buy a build-it-yourself piece of furniture from IKEA. The packing list tells you every single item that's in the box. But this ZTE packing list took it to another level—it not only spelled out every piece of ZTE telecom equipment in the box, but, more importantly, it listed ALL the US component-parts used in the construction of said equipment. Nine hundred seven glorious pages, detailing, in black-and-white, thousands of software and hardware component-parts manufactured by US companies such as Dell, Hewlett-Packard, Microsoft, Oracle, Juniper Networks, Cisco Systems, and Symantec. All being shipped to Iran as part of a telecom system that ZTE had sold the country.

Think about how bad this packing list was for ZTE. It didn't just show, for example, "one ZTE-manufactured spying telecom tower"; it showed "one ZTE-manufactured spying telecom tower containing a Dell widget, a Hewlett-Packard gizmo, a Microsoft thingamajig, and an Oracle thingamabob." Bad. Real bad.

Almost the minute the article hit, phones started ringing all around the building. Voices began shouting from offices that were habitually silent. Groups of employees in twos and threes began bustling by in all directions, engaged in fevered conversations. Suddenly the whole office was at DEFCON 2.

Everyone was in a panic, and they all seemed to be worried about one issue: how did Reuters get hold of this information? Where was the leak? Who had squealed? Why had they done so? I didn't understand why people were focused on this question since it was already too late to do anything about it. The horses were already out of the barn. Why worry about how the barn door got left open? The issue we should have been focused on was, "What are we going to do next?"

I finally pulled Meghan aside and asked her about this point directly. "Why are you—why is everyone—so concerned about *how* the story got out?"

And I'll never forget this: With a stone-cold glare, she looked me straight in the eyes and said, "Because now we can't hide anything!" And she scurried off down the hall.

Hide anything? Wait, what? What was she saying here?

This was more than a red flag; this was a piano dropped from a 747. If she was saying what I thought she was saying, then this would be a major crisis for ZTE. And it would certainly be a turning point in my ZTE career. Was ZTE knowingly involved in illegal activity and deliberately hiding it from US authorities? Holy crap. The situation would be bad enough if ZTE had somehow misinterpreted US embargo law; it would be exponentially worse if ZTE was breaking the embargo deliberately and hiding its activities.

If a cover-up was going on, how could I not have known about it?

Even though I wasn't personally responsible for any illegal activity—if, indeed, such was happening—and even though the allegations were about ZTE China, not ZTE USA, I still felt a growing sense of fear in my gut.

Donna spent the evening trying to talk me down from the ceiling and assuring me that everything was going to be okay. "You just need to do your job and try to protect your client as well as you can," she said, adding sarcastically, "That's why they're paying you the big bucks." We both had to laugh at that.

The very next morning I was working in my office—I believe I was writing another email, trying to light a fire under China to hire a damn law firm already—when I got a call from the receptionist.

"There's a gentleman here from the US Department of Commerce who would like to speak to you."

I went out to the reception area where a man in a grey suit stepped up to me and said, "Are you Ashley Yablon?"

"I am," I replied.

"I'm Agent Fulmer from the Department of Commerce, and I am serving you with this subpoena."

I accepted the envelope.

"Have a nice day, Mr. Yablon."

He left, and I turned and walked calmly back to my office.

Before even sitting down, I ripped into the envelope. The subpoena was calling for ZTE USA to produce two critical documents. Unsurprisingly, those two documents were: (1) the 907-page packing list that the Reuters article had cited; and (2) the accompanying contract that was signed between ZTE China and Iran for the $130 million deal.

I won't bore you with a ton of legal detail, but a key aspect of this development, and one of my predicaments, was that the US Department of Commerce knew it had no right or legal way to obtain those documents from ZTE China. In the legal world, we call this "lacking jurisdiction." The US government lacked jurisdiction over ZTE China, so it tried to circumvent that hurdle by asking ZTE USA to provide the documents by way of a subpoena. Their problem (and the US government was well-aware of this) was that ZTE USA was under no obligation or duty to provide such documents from China—even if subpoenaed. So the subpoena was merely a scare tactic. A very public one. An attempt to sink its claws into any part of ZTE it could get hold of, whether permitted to or not. But what this tactic also did was to put me, as ZTE USA's General Counsel, smack-dab in the middle of the mess with ZTE China, the US government, and eventually the Chinese government.

I immediately called Matthew, Meghan, and Zhang into my office for an emergency powwow. This subpoena thing was huge, I told them, and China needed to know about it pronto. We decided to call Mr. Guong One in Shenzhen and wake him out of his sleep. We scanned the subpoena and emailed it to him.

Guong One was calm, but I could sense turbulence below his stoic voice. "Thank you for bringing this to my attention," he said. "You did the right thing to call me. I will notify the appropriate people over here. Wait for further instructions from me."

As soon as I hung up, I sat down and wrote a companywide email and sent it out to all employees of ZTE USA, instructing them NOT to destroy any records pertaining to exporting. This kind of notice is a legal requirement for in-house attorneys. It's called a "litigation preservation" email. It puts employees on notice of pending litigation/investigation and forbids them from engaging in "spoliation of evidence." As I said, neither ZTE USA nor any of its employees were the true target of the subpoena (and no US staffers likely had any knowledge of the deal or any relevant documents), but I still had to put everyone on notice, given the fact that the subpoena was addressed to me and ZTE USA.

We mobilize

The Reuters piece and the subpoena finally spurred the Chinese bosses into action. Though these new allegations were not directly related to the HPSCI investigation, it was clear that the temperature had just been turned up to a boil in ZTE vs. US government relations.

Within twenty-four hours after the subpoena landed, I received a call from Guong One's office telling me that an executive team would be flying to Washington, DC, to personally conduct interviews with my five chosen law firms. The visiting team would include Guong One, Guong Two, the ZTE Chinese CEO, and a couple of other top people from ZTE China. Together with Xi, Len, and one or two others from the US side, the executive team would conduct the interviews as a group. Interestingly enough, they did not tell me what hotel they would be staying at. I'm sure it was nothing like my luxurious digs in Shenzhen, if such a place even existed in DC.

"Great," I said to Mr. Guong, "Let me know which days you want to do the interviews and who you want to talk to first, and I'll start coordinating things from my end and meet you all in Washington."

"That won't be necessary, Mr. Yablon," said Guong. "We will take things from here."

"Excuse me?" I said.

"You won't be involved with the selection process from this point forward," he replied.

"What are you saying?" My blunt demeanor surely fell outside of proper Chinese etiquette.

"I am saying thank you, Mr. Yablon, but your work is done for now. Perhaps we will see you when we are in Dallas and enjoy a cup of tea together."

I couldn't believe what I was hearing. They were going to freeze me out of the interview and decision-making process? Toward what end? This made no sense. I was the one who had done all the research and legwork to find these firms. I was the one who'd been talking to the attorneys and developing relationships with them. I was the one who would need to work with them, day in and day out, as the HPSCI investigation and Department of Commerce subpoena issue unfolded. I was the one who was going to be overseeing their work. Presumably.

And, oh yeah, I was also the only one in the company who knew a damn thing about US law. Why wouldn't they want me involved in the interviews—to ask the probing legal questions, to watch out for red flags? For the dozenth time since October, I found myself asking, *Why am I even here at ZTE USA? What's my role? Why am I constantly being shunted to the kids' table at Thanksgiving dinner?*

The executives arrived from China in Washington, DC, and, with an air of quiet urgency, conducted their closed-door conferences with the five law firms over a two-day period. Without me. Afterwards, they informed me that, against my advice, they had chosen to hire DD&M, the most expensive and—to my mind—least helpful of the five firms. This was yet another slap in the face. If they'd allowed me to participate in the interviews, I might have been able to talk them into a better decision. But no. They picked DD&M—largely, I suspected, because of Xi's persuasion.

My professional humiliation wasn't over yet. The executive team then tasked me with the job of reviewing DD&M's representation agreement and redlining the parts I thought were problematic.

The paperwork arrived, and I quickly went to work on it. A representation agreement outlines the terms by which a law firm will represent a client in a given legal matter. In this particular case, while the contract was officially between DD&M and ZTE USA, there was no mistake as to who was really calling the shots on the actual representation (and who eventually would be paying for DD&M's services): ZTE China.

DD&M's proposed agreement was, as I might have predicted, a middle finger to the very people who were hiring them. Excessive in every way. It stipulated, for example, that everyone from DD&M, including secretaries and paralegals, would fly first-class when traveling abroad and that all of their accommodations were to be only at five-star hotels. All meals were to be covered, with no expense limits. Multiple partners were to be hired to do the work, along with dozens of associates and support-staff members—all working simultaneously.

The representation agreement was basically a license to charge ZTE a king's ransom. Partners were charging $990 an hour, associate attorneys billing out at $650 an hour, and paralegals and support staff at well above normal hourly rates, even for a huge, top-flight firm. As written, the deal would cost the company hundreds of thousands, if not millions, of dollars per month. It was ridiculous and completely overblown. So I went to town with my red pen, marking, deleting, and changing DD&M's proposed terms. *No, you won't be flying first class. No, you won't be staying at five-star hotels; you'll be sticking to ZTE's travel policies like the rest of us. No, you can't have x number of partners and associates on our clock all at the same time. Yes, you must get prior written approval before incurring large expenses.*

Taking out the "fat" is what I do for any and all contracts that I negotiate on behalf of a client. This representation agreement was no different, and when I finished I felt good about what I had done to save the company money. A funeral home oughtn't take advantage of a grieving family during its most vulnerable time—that was my philosophy. I sent the redlined representation agreement to Guong One, Guong Two, Xi, Len, and a few other ZTE executives, letting them know exactly what I thought of it.

Their response? Leave it as is, no changes. Just sign it.

"But you told me to . . . "

"Just sign it."

So I did. Not a single edit was made to DD&M's original representation agreement. So much for trying to save my client money. DD&M would be making more in a week than I earned in an entire year. The whole thing made me feel physically sick, but at ZTE's insistence, I signed the agreement.

DD&M digs in

DD&M dug into its work—i.e., its billable hours—with a vengeance. They threw six or seven partners at the job, at nearly $1000 an hour, in addition to dozens of associates, paralegals, legal secretaries, and other support staff. That meant, basically, that every time they walked into the room, sat down, and opened their laptops, it cost us a thousand dollars. Every time they had a group conversation in the hallway, three grand. A full team meeting? Twenty-five thousand. You get the picture. That's how these big-time law firms work: efficiency and cost savings are words never to be uttered and are certainly not practices ever to be followed.

The legal team now had a dual track to its work. It still had to focus on the HPSCI investigation, as originally discussed, but now it also had to deal with the fallout from the Iranian deal that the Reuters piece had uncovered. Both were important tracks, but really they were separate ones, at least for now. The initial focus of the team's work was on the HPSCI investigation. First things first. DD&M quickly threw one of its big lobbyists at the case. This was a gentleman who had worked with House Intelligence before and had some connections there. He reached out to his contact on the House committee.

As soon as the committee got wind of the fact that DD&M had been hired, it moved into high gear. It began requesting a barrage of documents from ZTE. The committee's queries were mainly along two lines as reflected in the report quoted earlier: (1) it wanted documented assurances that ZTE's technology was safe and did not contain "backdoors" or the like (which was going to be a tough thing for us to prove, frankly; how do you prove a negative?), and (2) it wanted proof that ZTE was not, in fact, owned by the Chinese government.

It was the latter issue that seemed to spark the biggest concern. What the committee really wanted to know was: how can we be sure the Chinese government will not misuse technology that becomes embedded in American life, government, and public utility systems?

Ownership of the company was thus a major issue.

A brief word about this topic because it's crucial to the story. Business ownership in China is a thorny area, especially when it comes to large businesses. The bottom line is, if you're talking about a large Chinese business with a multinational reach—one that has access to major institutional funding and is in a competitive position with regard to regulations, market access, and other matters—then you can be pretty damn sure that business is in bed with the Chinese government. In the US, we like to talk about POEs (privately owned enterprises) and SOEs (state-owned enterprises) in China as if there is a clear difference between the two. We like to imagine the Chinese government exercises heavy control over SOEs, while allowing POEs to function fairly independently. In reality, SOEs and large POEs are virtually indistinguishable in most major ways. Both of them get help from the Chinese government, and both of them give help to the government by carrying out its policies. In other words, regardless of the technicalities of ownership, once you have reached a certain size and influence in China, it is because you have cultivated the right two-way relationship with state power.

The whole question of "Is ZTE privately owned or state-owned?" was not as meaningful, in my humble opinion, as the House committee seemed to believe it was. No matter what our documents might show, the fact was that ZTE was enmeshed with the Chinese state; there was no getting around it. Still, I guess the House committee had to look for something in its search, so this is the area where it focused much of its attention.

As I mentioned, the US government was looking at both ZTE and Huawei in this regard. Interestingly, ownership was an area where ZTE and Huawei differed starkly. Huawei's attitude was basically, "Yeah, we're state-owned. So what?" They didn't bother trying to hide it. ZTE, on the other hand, bent over backwards to prove it was not government owned. Its official position was that it was primarily owned by four private individuals/entities, and that the Chinese government held only a minority interest in the company.

That was the story ZTE was keen to tell. But, of course, it wanted to tell it in its own way, with its own chosen set of facts. It was not particularly jazzed about having the US government digging into its files and asking inconvenient questions. And it didn't really need to cooperate with the

US if it didn't want to. After all, the US government had no jurisdiction in China. So, although ZTE wanted to keep good relations with the US for many reasons, it wasn't about to give the US anything that made it look bad as a company.

In the early days of the investigation, Guong One's office in China was helpfully sending over many of the requested documents. I think ZTE wanted to be seen as cooperative; I also think it wanted to feed the investigators evidence that supported its claim that it was a privately owned company. But I began to realize pretty early on that there were potential problems brewing here. ZTE USA couldn't continue to be used as a conduit for prying information out of ZTE China. And I knew there was going to be a big problem when it came to the Commerce Department trying to use ZTE USA to extract that packing list and contract from Shenzhen. That just wasn't going to fly, in my mind.

One day I sat down with Jeff Pennington, the DD&M attorney in charge of the ZTE account. Jeff was a charismatic, glad-handing gentleman, as his role of head lobbying attorney for the largest lobbying firm in the world would imply. He kept that huge smile plastered across his face regardless of the problem at hand.

"We've got an issue shaping up, Jeff," I told him.

"How so?" said Pennington.

"Well, right now your firm is only representing ZTE USA. We're going to need a different firm to represent ZTE China."

"Why's that?"

"Because there is a conflict of interest here. Or at least a strong potential for one."

"I don't see it," said Pennington.

"Come on, Jeff. It's obvious. The interests of ZTE China and ZTE USA are not fully aligned. For example, there could be certain documents ZTE China might not want the ZTE USA office to see or to hand over, and so forth. What's good for the parent isn't always good for the child, and vice versa."

"I don't think there's a clear-cut conflict here," Pennington opined. "I don't see any reason DD&M can't represent both entities."

"I've just given you a reason," I pointed out. To me, the conflict was clear as day. Of course Jeff couldn't see it; all I imagined he could see were dollar signs for his firm and the swelling of his year-end bonus.

I tried to explain the concept of conflict of interest to my Chinese bosses, but as with other matters of US law, they claimed not to grasp the notion or why it was important. In fact, I believe they saw the issue very plainly (which would be a telling point later on).

And of course, DD&M was whispering in their ears: *Don't worry about it, we can handle both companies.* So it came as no surprise when the decision was made to retain DD&M as the attorneys for both ZTE China and ZTE USA. I knew this was wrong, and I felt it was a sleazy move on the part of DD&M. Even worse, I felt it was going to come back to bite us in the ass.

But once again I was overruled.

CHAPTER 8

The Heart of the Matter

In the days immediately after the Reuters article and the subpoena dropped, DD&M began ramping up its team, adding more lawyers, paralegals, and others. After all, DD&M was representing not just one corporation now but two: ZTE USA and ZTE China. They were going full-steam in their Washington office (their billable hours were going full-steam too). I still couldn't believe ZTE was giving them total carte blanche, but it was their money, I supposed, not mine.

For clarity's sake, here's a quick chart of everything in play so far:

Issue	Government Agency Waging "Attack"	What They Sought from ZTE	Legal Counsel Hired by ZTE to Defend
ZTE poses a threat to US National Security.	US House Permanent Select Committee on Intelligence ("House Intelligence Committee" or "HPSCI")	1. Assurance that ZTE didn't pose a threat. 2. Documents, answers to assure the US that ZTE's equipment did not have "backdoors," spying technology ("spyware") 3. Proof that ZTE was not owned/controlled by the Chinese government	DD&M – lead attorney, Jeff Pennington
1. ZTE is selling telecom equipment and spying technology to the embargoed country of Iran. 2. ZTE is buying US component-parts, putting them into their equipment, and selling it to embargoed Iran, which is against US law.	US Department of Commerce Agent Fulmer (server of the subpoena)	1. The actual contract between ZTE China and Iran 2. The 907-page "Packing List" documenting the spying equipment ZTE delivered to Iran, which included US-manufactured component parts	DD&M – lead attorney, Ricardo

And since I keep talking about ZTE China and ZTE USA, I thought it might make sense to list some of the key aspects that differentiate the two:

	ZTE China	ZTE USA
Location	Shenzhen, China	Richardson, Texas
Affiliation	- The "parent" company of ZTE worldwide - Headquarters for ZTE worldwide	- A "child" company of ZTE worldwide - Headquarters of ZTE USA
About the Office	- 30,000+ employees - Huge campus with its own bus system and hotel	- 300+ employees - Two floors of an office building
General Counsel	Mr. Guong (Guong One)	Ashley Yablon (me)
What It Does	Manufacturing hub of all equipment (phones, telecom equipment) sold worldwide (including the US)	Seller of ZTE China equipment in the US

As for me, I was trying to function as a liaison between ZTE, DD&M, the House Intelligence Committee (HPSCI), and the Department of Commerce, and putting out fires that were erupting in the wake of the Reuters piece.

One of the main fires was the barrage of emails, phone calls, and other correspondence I was receiving from all the US component-part companies that had been flagged in the Reuters piece. These firms, understandably, were freaking out. Suddenly they found themselves in the thick of an international scandal, and they were not relishing the press or the legal attention. Keep in mind, their contracts were with ZTE China, not ZTE USA. I had neither dealt nor negotiated with any of these companies.

So they were kind of barking up the wrong tree by coming to me. But regardless, each of these tech executives were saying to me, in one form or fashion, "What the hell are you up to, ZTE? We had no idea you were selling our components to banned nations. We are beyond shocked."

In my mind I kept wishing I could write back something like, *So let me make sure I understand this, Mr. US Component-Part Manufacturer: you sell hundreds of millions of dollars of parts to some Chinese company—and only now are you questioning where it goes and how it is used?*

The outrage was understandable but, frankly, a bit disingenuous. The reality—in my humble opinion—was that most of them knew damn well, or should have known, where their equipment was going when they sold it to China. This wasn't toothpaste they were selling after all, but highly specialized technology. But now that the shit was hitting the fan, they were all innocence and righteousness. Really?

They kept pointing out to me—correctly, alas—that in their contract with ZTE, the buyer of the equipment (ZTE) specifically agreed not to re-export any of the stuff to banned nations, even by building it into larger products or systems. In a contract, this is called the "export compliance" clause. Again, I did not negotiate ZTE China's agreements (just ZTE US's agreements), and hence I can't quote the exact wording of the export compliance section in the deals between ZTE China and these US manufacturers, but here's how a typical export compliance clause reads:

> **Export Compliance.** Purchaser understands that the Arms Export Control Act (AECA), including its implementing International Traffic In Arms Regulations (ITAR), and the Export Administration Act (EAA), including its Export Administration Regulations (EAR), are some (but not all) of the laws and regulations that comprise the U.S. export laws and regulations. Purchaser further understands that the U.S. export laws and regulations include (but are not limited to): (a) ITAR and EAR product/service/data-specific requirements; (b) ITAR and EAR ultimate

destination-specific requirements; (c) ITAR and EAR end user-specific requirements; (d) Foreign Corrupt Practices Act; and (e) anti-boycott laws and regulations. Purchaser will comply with all then-current applicable export laws and regulations of the U.S. Government (and other applicable U.S. laws and regulations) pertaining to the Products (including any associated products, items, articles, computer software, media, services, technical data, and other information). Purchaser certifies that it will not, directly or indirectly, export (including any deemed export), nor re-export (including any deemed re-export) the Products (including any associated products, items, articles, computer software, media, services, technical data, and other information) in violation of applicable U.S. laws and regulations. Purchaser will include a provision in its agreements, substantially similar to this Section, with its Sublicensees, third party wholesalers and distributors, who purchase the Products, requiring that these parties comply with all then-current applicable U.S. export laws and regulations and other applicable U.S. laws and regulations.

Clearer now? No, I didn't think so.

These US component-parts manufacturers were arguing that their asses were covered by this export compliance clause, so if ZTE had turned around and done something illegal with the parts, that was ZTE's doing, not theirs.

In reality, the issue wasn't quite so cut-and-dried, and they knew it. International trade regulations also state that a seller (e.g., a US component-part manufacturer) can't export any goods, tech, or services if the seller believes, or *has any reason to believe*, that such items are intended for re-export to a banned nation. And don't try to tell me these guys didn't have any reason to believe their items were being re-exported. So this show of outrage, I knew, was really a song-and-dance intended to distance themselves from the whole mess, and to show they had no idea (wink, wink) their goods would be used for nefarious reasons. But still, it was someone's job to field these protests, and since these companies didn't seem to want to communicate with ZTE China (their contractual partner), I, as the ZTE USA General Counsel, got to take the brunt of their anger.

We want a meeting

The US government, meanwhile, was starting to run into some brick walls in its demands for documents. DD&M was now telling the Commerce Department that ZTE USA would no longer be used as a vehicle to force ZTE China to hand over the documents demanded by the subpoena and that the Department of Commerce had no jurisdiction over ZTE China.

Commerce's response was, in effect, "Oh yeah, well we have jurisdiction over ZTE USA, so we're going to keep piling on the subpoenas, the searches, and the seizures, and basically making your life miserable until ZTE China cooperates."

To which our response was: Do what you need to do; it's not going to happen.

Finally, HPSCI grew exasperated with the whole tussle. It announced that it wanted to schedule some in-person meetings, in China, with both ZTE and Huawei. And it wanted to see the packing list and the contract when it was there. HPSCI also still had a host of questions it wanted answered—about the security of ZTE's tech and the nature of ZTE's ownership—and it didn't feel it was receiving the answers it needed.

So now we (DD&M, Meghan, Matthew, Zhang, and I) were going to pack up and take our show to China to meet with the HPSCI over there. Once in Shenzhen we would be joined by the ZTE China legal team—Guong One, Guong Two, and others, as well as key ZTE China executives, including the CEO of ZTE worldwide.

Huawei and ZTE, would, of course, be interviewed entirely separately, but both meetings would both take place on the same government trip, since both companies were located in Shenzhen.

We weren't thrilled about HPSCI demanding these meetings, but in a way the meetings were a good thing. They would give us something concrete to focus our energy on, and we hoped the outcome would be a quick and favorable resolution of most of the issues. HPSCI didn't give us much time to get ready, so we dove into our prep work immediately. DD&M flew to Dallas with its full team, and we buckled down and got to work. This entailed: (1) formulating our game plan regarding answering HPSCI's questions; (2) determining who would be the "emcee" and lead the tour on the ZTE factory floor; (3) finding and producing relevant documents that addressed HPSCI's litany of questions; and (4) figuring out the best people to meet with HPSCI and prepping those people with specific answers to likely questions. It was less than two weeks before we headed to China, and we had a load of work to do.

In the meantime, besides answering the barrage of hate mail from US component-part manufacturers, dealing with the HPSCI investigation, handling the Department of Commerce subpoena, and prepping for the HPSCI meeting" in China, I still had to find time to do my day job of reviewing sales contracts, managing litigations, overseeing employment matters, and more. Busy time indeed.

To China again

I decided to travel to China on April 6—Good Friday—ahead of most of the DD&M team (Meghan, Matthew, and Zhang had already preceded me). I missed Easter with Donna and my family. This was to be my third or fourth trip to Shenzhen, so I knew the ropes and was resigned to the travel conditions. Donna knew the drill too and made sure I packed tons of snacks to compensate for my lack of a meal budget: almonds, dried fruit, and Power Bars. "Be careful, baby," she said. Prescient words, it would turn out.

The plan was that a few of the DD&M support staff (paralegals) would travel with me, and the rest of the DD&M group (i.e., the $990/hour partners) would arrive in China after Easter. Hey, don't let a little government investigation in China get in the way of your holiday plans. That was insulting, for sure, but nothing could have prepared me for the insult I was about to experience during the travel itself:

Picture me juggling my luggage, computer bag, and armload of company documents as I jostle my way to the back of the plane amongst the final boarding group. Lack of overhead space by this point a given. Crammed for the next twenty hours into a tiny middle seat between two large individuals.

And whom did I need to pass in order to get to my spacious quarters? None other than the DD&M paralegals: first to board, now sitting in first class with drinks already in hand, watching me fight my way to the back, offering only a condescending head nod to acknowledge my existence, checking their e-mails and texts in their cozy, extra-wide seats.

The crazy dichotomy between DD&M's travel accommodations and mine continued throughout the whole trip, by the way. DD&M stayed at the Shenzhen Ritz-Carlton or some other fabulous five-star hotel, while I stayed at Meet at The Blue Sky in—believe it or not—the Winnie-the-Pooh-room again. I envisioned them beginning their day with a work-out in the state-of-the-art in-house gym, a sauna, a whirlpool bath, and a Michelin-quality breakfast in the tastefully lighted dining room. I began mine with a backache and a Power Bar. They went out to lunch at local gourmet restaurants, where they ordered $75-a-bowl shark fin soup and put it on the ZTE tab. I ate my second Power Bar of the day on the Hospitality floor. In the evening, they embarked on their serious dining excursions—complete with fifteen-year-old single-malt scotch, no doubt—while I grabbed a value meal at KFC. And to add more salt to the wound, the DD&M people never once invited me out to dinner, even though I had procured them the job that was keeping the little mints on their pillows. So much for professional courtesy.

The ZTE brass only perpetuated this indignity. They treated DD&M like the queen bee while they buzzed around the hive like worker drones, trying to make life as comfortable as possible for the silk-stocking law firm. Meanwhile, I would encounter the Great Wall of China every time I asked for the least thing from ZTE or tried to solve their problems.

Bitter? Yes, I was; I'll admit it. I was the GC who had worked his ass off to get this in-house gig. I was the guy who had hired DD&M and who'd agreed to their ridiculous demands. And their thanks to me was to take all the ZTE China and ZTE USA execs (including Meghan, Matthew, and Zhang) to the finest restaurants in Shenzhen while I ate "Chizza" at KFC. Pretty funny when you thought about it: DD&M taking ZTE out for meals that ultimately ZTE paid for.

Back to the business at hand. As I waited for the rest of the DD&M team to arrive in Shenzhen, I made good use of my time. I scheduled my first meeting with Mr. Tang, who headed up the Export Department for ZTE. I figured I should get a jump start on understanding the inner workings of the exporting of ZTE products. Surely the Reuters article couldn't be fully accurate. Nothing that nefarious could be happening at my company, right?

Mr. Tang explained to me very matter-of-factly that ZTE owned a number of sub-companies that bought US-made telecom products from US manufacturers. He outlined the sub-company structure as a strategy the company used for business and tax-related purposes. Holy crap. Here was the head of ZTE's worldwide Export Compliance Department cavalierly explaining to me a monumentally shady scheme of using "shell companies" to skirt regulations. His delivery was so deadpan I was literally waiting for him to stop and grab my arm as he broke out in peals of laughter, saying, "Man, I'm just yanking your chain! That would be sooo illegal, right?"

But that didn't happen.

I was starting to see, for the first time, the outlines of something I really didn't like, but I didn't want to jump to any conclusions.

When the full DD&M team arrived, we held a whole-staff kickoff meeting in the biggest conference room on the ZTE campus. The DD&M lawyers took the lead, while Guong's office provided logistical and organizational support, helping the lawyers find whatever they needed to find (well, whatever ZTE wanted them to find). The agenda of the kick-off was basically, "Here's who we'll be meeting with and here's what we need to present and prepare for them."

DD&M's attorneys and staffers worked alongside ZTE China's in-house attorneys, each assigned tasks such as creating slides and organizing documents. As I mentioned before, there was to be a hard-hat walk-through of the manufacturing facilities when HPSCI arrived, as well as tightly scheduled presentations of various materials. The plan for our day one was for the legal team to get busy with their presentation materials and then reconvene to do a test run of the whole event the next afternoon. The group broke up, and everyone was given comfortable office space in which to work.

Well, everyone on the Track One team, that is. As you'll recall, there were two main tracks for these Shenzhen meetings. Track One was focused on the HPSCI investigation, and Track Two was focused on the Commerce Department's subpoena. The vast majority of the legal team's energy—and over ninety-five percent of its manpower and resources—was allocated to the House Intelligence stuff, Track One. Track Two, the Commerce Department's subpoena, was to be handled, it turned out, by me and one attorney from DD&M, a guy by the name of Nueve.

Guong Two was directing traffic after the big meeting, telling everyone where to go and whom to see. He directed Nueve and me to a run-down cubicle that was to serve as our temporary work space. On day one, Nueve and I had a series of meetings and work sessions that weren't particularly revelatory. But on day two, that all changed. In a big way. Guong Two told me and Nueve to go upstairs and meet with Preston; he was expecting us.

You remember Preston—he of the weird December phone calls.

The big reveal

The moment Nueve and I stepped off the elevator on floor sixteen, it seemed as if we had entered an alternate universe. There was no air conditioning, and the uncirculated air had a stale and dingy quality. The whole ambience of the floor felt untended and frayed around the edges, like an old bus or an airplane that should have been retired years ago. This was clearly a section of the building where guests were never meant to be taken. Not a single office seemed occupied. Old retired chairs were stacked up next to empty rooms. Dust was collecting everywhere, and outdated office equipment lined the walls. It looked like the hallway you walked through in a horror movie. Were they trying to send me a message?

"Welcome to the real ZTE," I half-joked.

Nueve and I walked down the poorly lit hallway and knocked on the door of the small conference room to which we'd been directed.

"Enter, please," came a voice from within.

As we stepped into the room, I felt as if we were stepping onto the set of a B-grade spy thriller. The room was bathed in darkness—no windows, no lights on, only what little ambient light was leaking in from the hall. A man sat at the end of a table, his face obscured by shadows. A small banker's lamp, turned off, and a laptop computer sat in front of him on the table.

"Please, have a seat," the man said in an unnaturally deep voice, almost as if he was trying to play to the thriller-movie vibe. It was Preston; I recognized his voice and broken English.

I laughed, as I am prone do when I'm feeling uncomfortable. "Any chance we can have a little light in here?" I said, looking around for a wall switch.

Preston did not answer me right away. Instead, he sat there in pointed silence for several seconds.

"Here, let me turn this on," he finally said, pulling the cord on the small table lamp. The dim light cast Preston's features into creepy relief, doing little to dispel the ominous atmosphere.

Nueve and I sat at the table. Silence reigned for an awkwardly long time.

I took it upon myself to break the ice. "So hello, Preston," I said. "I'm sure you've seen the subpoena, so you know why we're here—to see the packing list and the contract, and to get our copies."

A long pause ensued.

"I'm afraid that won't be happening today," Preston replied.

"What do you mean?"

"You will not be receiving hard or soft copies of the documents. You may look at the electronic copies for fifteen minutes each, but you may not take them into your possession."

"We're the attorneys on the case," I reminded him. "We need to prepare for the Commerce Department's request. How are we supposed to do our job if we don't have the documents to study and analyze?"

Think about that. We're paying this elite law firm to fly all the way to China from Washington, DC, paying them millions of dollars—and yet we're being given fifteen minutes with the most important documents in the case?

"Those are the conditions," replied Preston. "Do you agree, yes or no?"

"Fine," I said. I'd have to wrestle with Guong One later on to get my copies. For now, I just wanted to see the documents. "Let's have a look at them."

We waited for Preston to make the next move. He remained still for a moment, then shut the lamp off, plunging the room into darkness again. There was another long pause. Was he deliberately screwing with us?

"Please turn off your cell phones and place them on the table, face down," he instructed us.

He had to be kidding. Was he about to commence an inquisition? Or was he just worried we might snap photos of the documents? I sighed my annoyance and complied with his request. Nueve did too. Did he know something I didn't?

Only when Preston was satisfied that our phones were safely disabled did he wake up his laptop, throwing a little light back into the room. He then turned on a small projector I hadn't noticed. The image of his laptop screen was cast onto the wall.

So this was how we were going to review the key documents? By looking at shadows on a wall? So be it. I pulled out my notebook and pen; Preston didn't seem to object to that.

Nueve and I stood up so we could be closer to the projected image. The first document Preston was showing us, apparently, was the packing list. In the upper corner it read "page 1 of 907." Preston began slowly scrolling through the list for our benefit. What we were seeing were clearly the components of a cell-phone tower system that had been sold to Iran, along with related equipment—routers, software products, phone equipment, networking devices.

"The Reuters article was right," I observed. "This stuff has the capacity to conduct heavy-duty surveillance."

Nueve nodded his agreement. "So how do we do this?" he said to me. "How do we use our time most efficiently? This document's 907 pages and we've only got fifteen minutes to look at it."

"Is this a searchable PDF?" I inquired of Preston.

"It is," he answered.

"Let's just do a search for each of the American tech companies," I suggested, "and see how many times their names pop up on the list."

Off the top of our heads, we started calling out the names of the US component-part manufacturers mentioned in the Reuters piece, one by one: Cisco Systems, Microsoft, Symantec, Juniper Networks. Preston would type the name into the search box and then scroll through the long document looking for highlighted items. Hundreds—sometimes thousands—of hits were coming up for each company's products. The list was exactly as described in the Reuters piece: documentary proof of a spy-capable cell-tower system that had been sold to Iran and contained thousands of components made in the US. It was black-and-white evidence of wrongdoing, whether criminally intentional or not. As I have mentioned several times, the US strictly forbade American-made components from being sold to Iran, regardless of whether they were sold separately or as part of a larger system.

I hastily scribbled as many notes as I could. My goal was to list out each US manufacturer and how many times it appeared in the packing list. With only fifteen minutes to work with, though, I managed to get through only a handful.

After those few searches of the document by company name, Preston abruptly stopped scrolling and clicked the document closed. Our fifteen minutes had flown by. So much for the packing list.

Nueve and I next asked Preston to show us the contract. As with the packing list, he projected the image of the contract on the wall.

And that was when my life changed.

In every story, there is a defining moment—a precise instant in time—when the game changes for good. The whole story, from that point forward, can be divided into a "before and after" scenario—things that happened before the critical event and things that happened after. I was about to experience the defining moment in this story, but, of course, I didn't know it.

The contract document was a long one, and initially it looked like any international contract I had seen before—divided in half vertically, with the English-language version on one side and the other country's language (in this case Chinese) on the other side. As Preston began scrolling through the contract, I was seeing several oddly named entities as the parties doing business. A couple of these were names I recognized from my earlier meeting with Mr. Tang. They didn't make immediate sense to me. I was also seeing the word "Iran," popping up with frequency throughout the contract.

Suddenly a group of words flashed by on the scrolling screen that struck my eye like a hot ember. I was sure I had read them wrong.

"Stop," I snapped at Preston. "Scroll back, please."

Preston rolled the document back two or three pages.

"Hold it there," I said. I'm sure you could hear my chin hit the floor in the silence of the room as I was rendered momentarily speechless. Right there in front of my eyes, sprawled across the screen in bold letters, were the words, "How We Will Get Around US Export Laws."

I groped for the back my chair. I needed to sit.

CHAPTER 9

The Words I Couldn't Unsee

I literally could not believe what I was seeing. The contract spelled out, in black and white, an official corporate intention on the part of ZTE to sell billions of dollars' worth of US equipment to US adversaries. In terms of evidence, this was like the police finding a note in the home of a murder suspect entitled, "How I Will Murder My Wife" with step-by-step plans. The contract couldn't have been clearer in its intentions or more brazen in its description of its actions.

Below the heading, "How We Will Get Around US Export Laws," was language explaining the plan in detail. In essence, a series of shell corporations was to be created, which would be used to disguise the buying and selling of the US-made components, much in the same way organized crime uses shell companies to carry out money laundering and other illegal activities. Below the brief description was a table with four separate boxes containing the names of ZTE and three shell companies, all owned by ZTE: ZTE, ZTE Kangxun Telecommunications, ZTE Parisian, and Beijing 8-Star. And in each box their roles were duly explained. ZTE Kangxun Telecommunications would buy the parts. Beijing 8-Star was designated to sign contracts at the Iranian end of the pipeline, buy the products, and re-export them to Iran; and ZTE Parisian provided engineering services (installation) to the customers.

I continued to take notes on what I was seeing, and Preston, for some reason, did not stop me. What we were looking at was a blueprint for how ZTE planned to work with not only Iran, but with numerous other embargoed countries, such as Sudan, North Korea, and Cuba, to generate billions of dollars' worth of global business in contravention of US law.

I was perspiring through my Uniform, and not just from the stuffy air.

After studying the screen for several minutes, I was in the middle of writing a word in my notebook when Preston shut the projector off.

"Okay, we're done," he said with military finality, clicking the little lamp back on.

I said, "I need this information, Preston, in order to prepare for the meeting with the Department of Commerce. The more I know, the better."

"We are done," he repeated, his shadowy face looking every inch like a movie villain's.

I knew we had seen all we were going to see of the docs. At least for now. "So what's going to happen when the US authorities demand to see the packing list and contract based upon their subpoena?" I asked Preston. "Are you going to tell them they can't have the documents either? Will you ask them to watch a fun slide show instead?"

"That remains to be decided, after the whole legal team has had a chance to consult on the matter. Is that all, gentlemen?"

"One more question," I said. "Where is all that equipment right now, all that stuff on the packing list?"

Preston paused, considering whether to answer me or not. "It is sitting in several dozen wooden crates, each about the size of this room, in a warehouse in Iran."

"Awaiting assembly and servicing by ZTE Parisian?"

"That is correct."

I jotted a final note in my notebook and made my exit. As Nueve and I stepped out into the hallway, I was literally shaking. I knew the importance of what I had just seen, and I knew I had just been burdened with information that was going to drastically alter my career—and my life—one way or another, depending on what I did or didn't do with it. A line had been drawn in the sand, but I didn't want to look at it. At least not immediately. It felt like too much to digest at once.

"What now?" I said to Nueve, struggling to keep my voice light and casual. "Try to reconnect with the rest of the team downstairs?"

"Guess so," said Nueve, and we headed silently toward the elevator. I knew he, too, was aware of the significance of those documents we had just seen. As a lawyer, he had some decisions to make as well. But unlike me, he didn't have anything to lose personally; he was just a hired gun. He would be able to sleep just fine tonight, and not just because of the sateen sheets on the king-size bed at his five-star hotel.

I'll never forget this moment as long as I live: As we were stepping toward the elevator, Nueve stopped in his tracks and looked at me and I felt as if his eyes were fixated on my notepad full of notes still in my hand.

The pause felt like an eternity.

What I gleaned from his silence in that eternal moment was this: "You might want to lose those notes before going back to the States."

I remained frozen in place as Nueve walked ahead of me and stepped into the elevator. The doors closed and I let him descend without me.

"What if we say...?"

I had more meetings and tasks to attend to that day, which I did with a heavy spirit and a distracted mind. As I rejoined the legal team downstairs later in the day, I was still ruminating on what I had seen in that darkened room, and on Nueve's ominous "advice" to me. Pennington, as always, seemed upbeat and optimistic. His attitude helped to lift my spirits a bit. I asked him how the preparations for the HPSCI meeting were going.

"Superbly," he told me. "We're ahead of schedule. We've found some excellent documents that will help make our case on the ownership front, and we've got a great English-speaking tech expert who's going to address the tech issues in a presentation and a Q&A session. All our slides and docs are looking good. We've got answers lined up for all the questions they sent us. We'll be more than ready for the House Committee tomorrow."

I explained to him what I had seen upstairs. It didn't seem to dampen his mood too much. "We may ultimately decide not to show those documents," he informed me. "They're discussing it in there." He nodded to a side room, where I saw several ZTE China in-house attorneys and executives, including Guong Two and Meghan, having an animated discussion. "We'll get back to you on how to handle the subpoena. Cheer up; it's all good."

It wasn't all good, not in my opinion. Not by a long shot.

I wandered over toward the room where the discussions were going on in Mandarin. Guong Two spotted me and hurried out of the room. Guong Two was not a large man but he did command attention whenever he entered a room—and knowing that, he always played that angle up. With a furrowed brow, he leaned in toward me and said in a hushed tone, "What if we say we never shipped any of that material to Iran?"

What? That was the level of discussion they were having in there? He wasn't seriously proposing this as a plan, was he?

"The cat is already out of the bag on that," I said to him. He looked at me oddly. *Speak literally,* I reminded myself. I rephrased my point: "It's already public information that the materials were shipped. We can't just say they weren't."

Guong Two considered this briefly, then hurried back into the room to talk to his colleagues some more. Heads nodded furiously. After a couple more minutes, Guong Two came back out and addressed me again. "What if we say the packing list Reuters acquired was not a real document? It was fake."

Again, was he really suggesting this as a strategy?

"First of all, it wasn't fake, it was real," I replied. "Second, we don't know who the source was who leaked the list to Reuters. It may have been someone within the company. It may have been someone who has plenty of evidence to show the list's authenticity. If that's the case, we would only be making matters worse, not better, by lying. That would be a dangerous route to take."

"What if the list didn't identify the US companies?" he pressed me.

"But it does identify them. I just saw it upstairs. Are you suggesting we alter the document by deleting the names? That won't change the copy of the document that's already out there. And besides, it wouldn't take much for someone to identify the manufacturers just by reading the product details."

Guong Two went back to the huddle and they talked some more. Out he came again.

"What if we go over to Iran, open the crates, and switch out all the American parts with non-US parts?"

Unbelievable. Each idea he was proposing was more illegal and out-rageous than the one before it. This idea had so many obvious flaws as to render it ludicrous on the face of it. But he was serious. Dead serious.

"We're talking about thousands and thousands of parts, in dozens of massive wooden crates," I said. "Think of the logistics! And where are you going to find all these new non-US parts? There's a reason ZTE uses US parts in the first place—because in many cases, the US is the only country that makes them. And anyway, this strategy still wouldn't change the fact that the packing list has already been publicized. There's no way of getting around that."

Guong Two returned to his group, talked some more, came back to me yet again.

"What if we go over to Iran, find all the US parts in the shipping crates, and scratch out the serial numbers?"

Wow. I stood there with my mouth wide open, dumbfounded that this discussion was even happening. Just as with my earlier conversation with Mr. Tang about the shell companies, I truly thought TV cameras must be recording my reaction before I was let in on the gag. But no, this was real.

"I will relay your opinion to the others," he said. He returned to his group and commenced talking again, leaving me shaking my head in disbelief. After a couple of minutes, Meghan stepped away from the group and approached me.

"Magic," she said at half-volume.

I nodded, looking around to make sure we were alone.

"Do you hear what they're saying in there?" she asked me.

"I hear them talking, Meghan, but I don't speak Mandarin, so no."

"They're saying they want to shred all the documents related to the subpoena, and wipe them from the servers and from everyone's hard drive. They're going to lie when the Commerce Department asks to see the docs. They're going to say we don't have them. They're thinking about going over to Iran and changing out all the parts with new ones."

"That's crazy. We're talking about thousands and thousands of parts."

"You forget. One thing we have in China is human capital. Manpower is not an issue."

"And how are they going to sell this crazy story to the Department of Commerce? Those people aren't idiots."

"They're not going to sell the story," Meghan said. "You are."

"What?"

"They still see you as the person who will stand up and tell the story to the authorities. They are planning for you to be the—what is the term— escape goat."

For the third time that day, my jaw dropped to my chest.

Meghan whispered, "Magic," turned, and rejoined the group.

My heart began racing. Within the span of twenty-four hours, I couldn't believe what I had discovered:

1. ZTE had set up an elaborate shell company scheme to buy US-component parts and sell them to embargoed countries.

2. The 907-page packing list was real and showed in great detail all the US component-part manufacturers and exactly how their products were embedded in ZTE telecom systems and equipment that had been sold to Iran.

3. The contract between ZTE and Iran had the audacity to spell out exactly how the shell-company plan would go down and to display charts showing the roles all the parties would play in the scheme.

4. ZTE planned to lie to HPSCI and to the Department of Commerce and refuse to comply with the subpoena. And . . .

5. The icing on the cake? ZTE wanted me to stand up and knowingly lie to Congress.

Was this really happening to me?

I spotted Pennington packing up his briefcase and made a beeline for him. "Do you know what's going on in there?" I hastily explained to him that ZTE was planning to lie to our government and that they wanted me to be their mouthpiece. Pennington pretty much blew me off, minimizing the whole thing. Sure, what did he have to lose? He was outside counsel—it was no skin off his back—and he was probably distracted at that moment, thinking about the five-star gourmet dinner he was soon to be enjoying courtesy of his fat ZTE expense account.

Sick call

I returned to the Winnie-the-Pooh room at Meet at The Blue Sky and proceeded to wear a groove in the cheap carpet with my back-and-forth pacing. The pressure in my chest had now turned to physical pain, and my heart felt like it was going to explode. I needed to talk to Donna, so I waited till I was sure she'd be awake, Dallas time, and called her from my room at the hostel, paying the ridiculous international charges.

After I explained to her everything that had happened and my fears of being made into the "escape goat," she said to me, "Ashley, you need to come home. We need to talk about this. We need to figure this out. Make an excuse. Leave early. Get your ass home."

I didn't need much convincing. Donna was the person I trusted the most on Earth, and I needed to be with her ASAP and figure this out. "Okay," I said, "I'll catch the next flight I can."

"Oh, but Ashley, be careful," she added. "I hear they've been stopping ZTE people at the airport."

"Great."

Life just kept getting better and better.

The next day was the dog-and-pony show for the House Intelligence Committee. That was DD&M's event, and I wasn't present for it. In the later part of the day, the whole legal team, including Guong One and Guong Two, gathered back in the large conference room to conduct a post-mortem of the meeting. There was much backslapping and a celebratory feeling in the air. Everyone seemed to feel things had gone quite well, better than could have been expected.

I was glad to hear the good news, and I hated to pull a Debbie Downer, but I still needed an answer to my nagging questions about the packing list and contract. But Pennington was too busy patting himself on the back, proclaiming the success of the walking tour, the Q&A session with the HPSCI team, and the elaborate slide show.

"This was a tremendous success," he said to the crowded table of nearly thirty ZTE China and ZTE USA attorneys, the full DD&M team, and several ZTE executives. The DD&M team all beamed as if they had accomplished something major. Guong One kept that pleasant, professorial smile on his face. Xi lost his puzzled expression and was leaning back in his chair grinning from ear to ear as if he had personally contributed to the good news. Guong Two retained his pensive demeanor—which terrified me, knowing what he was contemplating.

Finally, when there was a break in the verbal high-fives, I sought to address the elephant in the room that seemed to have gone invisible with all the success of the HPSCI meeting. "So how are we going to handle the subpoenaed documents?"

There was a palpable drop in the energy level of the room, and after all eyes shot me a laser-focused glare, they turned to Guong One for his response. He steepled his fingers, shook his head as if contemplating something huge, and looked slowly around the table. "We have given this matter careful consideration," he said at last, "and we have decided to comply fully with the US government's requests. We will give them the packing list and the contract. There will be serious consequences for the company as a result of this action, but we feel this is the only way forward."

There was a chorus of relieved *that's great, wonderful news,* and *yay team*-type reactions around the table. I felt a tremendous weight lift off my chest. At least I would be safe. At least I wouldn't be asked to lie. Yes, the company would be heading into stormy waters, but at least it was doing the right thing now. As a team, we would find a way to handle the fallout.

The meeting ended, and the group broke up into small, chatting clusters. The self-congratulations continued ad nauseam.

Now that the major issue of the subpoena materials had been resolved, I felt this would be a good time to make my getaway. I took Guong One and Pennington aside and told them, "Listen, I'm not feeling well—maybe it's the food I've been eating over at my 'hotel' (a little dig about my living quarters; why not?)—so I'm going to leave early. I think the rest of the team can handle things from here."

People were surprised by my decision but everyone seemed okay with it. I said my goodbyes and started off down the hall to collect my luggage and make my departure.

I walked with a little spring in my step back to my temp-employee cubicle and gathered my coat, my briefcase, and my luggage. I noticed a message blinking on my office phone and a couple of memos on my desk. They looked urgent. I ignored them and stepped out into the empty hallway.

As I pulled my suitcase down the hall, the clacking sound of wheels-on-tiles echoing off the walls, my thoughts of what I gleaned from

Nueve's silence played over and over again in my head. *"You might want to lose those notes before going back to the States. You might want to lose those notes before going back to the States. You might want to lose those notes . . . "*

Upon turning a corner and heading toward the lobby, I spotted a rounded, dome-top metal trashcan standing outside an office door. Its presence seemed fateful. My feet suddenly seemed to gain weight. I stopped in front of the receptacle. *"You might want to lose those notes. You might want to lose those notes . . . "*

I sighed. In a weirdly emotionless way, I opened my briefcase and removed my notebook. I stood there for what felt like a decade, trying to decide what to do. Take the notes with me or throw them in the trash?

I tore the pages out of the book and held them in my hand, wishing like hell there was a moral guidebook I could follow in a situation like this. *Tell me what to do*, I said silently to no one in particular. I received no answer.

I pushed the lid to the trashcan open, still clutching the notes as Nueve's silence rang through my head over and over. Finally, I opened my hand and heard the pages hit the bottom of the trashcan as its hatch snapped shut.

I'd only walked a few yards when a voice behind me whispered, "Magic."

Uh-oh.

I turned to see Meghan scurrying to catch up with me. "Ashley, before you leave, I need to tell you something. Mr. Guong Two just told me we're not going to comply with the subpoena. He says it's against Chinese law for us to provide such documents to a foreign government. Mr. Guong One was lying to everyone in the meeting just now. Thought you should know. Magic."

Fuck.

CHAPTER 10

Putting the Pieces Together

A twenty-hour flight gives you a lot of time to think. My return trip from Shenzhen to Dallas was one of the longest days of my life.

On an immediate level, I was worried about being stopped at DFW airport when I landed. Donna had told me they were detaining ZTE people. As a US citizen, would I be allowed to enter the country, or would I be refused entry? If the latter, what would happen then?

Would the authorities interrogate me? Which authorities? Would they search my luggage, my laptop? Detain me? I had no idea what to expect, but the added anxiety was doing nothing to help my mental state.

On a deeper level, my mind was busy assembling the pieces of the puzzle that had been staring me in the face almost from the moment I was first hired at ZTE. All the red flags I had been missing because of my naïveté and cheerful goodwill suddenly began waving at me in full crimson glory. I thought back to the Thanksgiving meeting and my bosses' stated hope that I would stand up and "testify on the Hill" for them. I recalled the bizarre phone and email exchange in which Preston was asking me how to legally sell to embargoed countries. I thought about my first trip to China and how worried everyone was about the investigation despite their insistence that ZTE was completely in the clear. I recalled the bizarre decision to cut me out of the hiring negotiations with the law firms. And, of course, I reflected on the newly learned fact that ZTE was planning to lie to the US investigators and refuse to cooperate with the Department of Commerce subpoena.

There was only one obvious conclusion to be drawn from all of these disparate threads, but I didn't want to jump to it. Not until I had processed all this stuff with my trusted confidant, Donna. I couldn't wait to get home.

If they would let me through Customs, that was.

I might have slept a few hours on that flight; I'm not sure. If I did, my dreams were filled with the same anxieties that were assaulting my waking mind.

Summit on the home front

"Thank God you made it," said Donna, greeting me with a hug on the home side of Customs. I had somehow cruised through the screening process without issue. "Let's get you home and rested, and we'll figure out what to do."

Donna and I spent the entire weekend trying to sort out the pieces and decide what to do next. "Meghan actually used the word 'scapegoat'?" she asked me, in one of our many talks.

"Well, she said 'escape goat,' but the meaning was clear. Those guys were planning to throw me under the bus over there. They were going to have me lie to the US government. But I'm starting to think the whole thing is more insidious than that."

"What do you mean?"

"I'm starting to think they've been setting me up to be the fall guy—I mean the big-time fall guy—since the day they hired me. Since before they hired me. I'm starting to think they've been viewing me as their patsy since day one."

"I'm sure that's not true, Ashley, is it?"

"Think about it. They cherry-picked me when I was still working at Huawei. I always wondered why they pursued me so openly in those early days. It seemed kind of weird. But looking back, I think I must have fit their perfect profile of someone they thought they could manipulate, control, and eventually compromise."

"You were also very qualified; don't forget that."

"But that's not what they were looking at. That's not what sold them on me. When they looked at me, they saw a young, ambitious lawyer who was hungry to land a job as General Counsel for a major corporation. They knew I had worked for Huawei and they probably figured I was okay with the things Huawei was doing. They thought I'd be so smitten with the idea of being lured away by their rival and offered a top dog position that I'd gladly wear blinders for them. Play the happy idiot. And you know what? They were kind of right."

"Oh, come on. Don't beat yourself up. There's no way you could have known you would end up in this position."

"I should have, though. I should have seen it. The signs were there. The red flags. I just didn't want to look at them. Think about the weird way they treated me from the get-go. They hire me to be their General Counsel, but then they exclude me from all the important conversations; they keep major things secret from me; they go against every recommendation I make to them. Why hire me at all if they're going to treat me that way?"

"I'll admit, that was pretty strange," Donna said.

"There's only one reason I can think of for them headhunting me, hiring me, and then keeping me in the dark. Because they had plans for me. They were grooming me to be the innocent fool who'll do as he's told because he's too blinded by his own ambition to see what's really going on."

"I hate them, Ashley."

"And you know what happens to innocent fools? They're the ones who end up going to prison.

"So what do we do?"

"The billion-dollar fucking question."

"You've got to quit," Donna insisted. "That's the only thing you can do. And once you're safely out of their clutches, you've got to go to the feds."

"Really? And how's that going to work out for us? What are we supposed to do for income? You're just trying to build your practice; we can't depend on your earnings yet. Have you looked at our student loans lately? Our mortgage payments? Your monthly trips to Colorado? We won't last two billing cycles without my salary."

"So you'll get another job."

"Where? What kind of reference do you think I'll get from ZTE? Who do you think will hire me knowing I walked out on my previous employer under a dark cloud? How thrilled do you think any company will be to hire a whistleblower as their next General Counsel?"

"We'll figure something out. We always do. You just need to get the hell out of there and away from them and their clutches. Those guys are dirty."

"Do you really think I can walk away at this point?"

"What do you mean?"

"I mean the government is all over this thing now. The packing list has already been published. The House investigation is just getting warmed up."

"So? That's not your problem."

"It is, Donna. It is my problem. I'm front and center in this whole thing. They know me now, the government people. They've talked to me. They know where you and I live. They know my role in the company. I can't just walk away; I'm in this, for better or worse. They're going to hound me wherever I go. They're not going to forget about me, regardless of whether I still work for ZTE or not. I'm their star fucking witness. I'm the one—the only one—who can talk about both ZTE and Huawei. I'm the fucking unicorn once again."

"But at least you'll be safe if you quit."

"Safe? You're joking, right? If I blow up a bunch of hundred-million–dollar high-tech deals with some of the shadiest governments on the planet, who do you think's going to want to sell me a life insurance policy? How much do you think my life will be worth?"

"Oh, come on," Donna responded, "it's not that bad. This isn't a Hollywood movie."

"It's worse. I've heard things I won't even tell you about the Chinese. They don't fuck around. This is serious, serious shit. This is make-a-phone-call-to-the-*Triad* shit. I'm not even kidding. This is Ashey-Yablon-disappears-from-the-face-of-the-planet-without-a-trace shit."

"So what are you saying? That there's no right course of action here? That there's nothing you can do that will change things for the better for us? That we're screwed whether you keep your job or quit it?"

"That sums it up pretty well, I'd say."

Lawyering up

Donna and I were both attorneys, but we had no idea what to do, how to proceed. As I've pointed out many times, law is a highly specialized profession, and most lawyers have no idea what to do when they get in legal jeopardy themselves. About the only thing Donna and I could agree on was that I was in big trouble and needed to talk to someone.

The first person I thought of was Robert Winston. Robert was one of my best friends. I'd met him at one of the firms where I worked when I was "rounding out my tool belt." I had actually helped him get a job at McAfee before I was hired there, and he was the one who helped me get my job there. Robert was a terrific guy and brilliant in a think-outside-the-box kind of way. He was a legal MacGyver-type who always knew how to maneuver around tricky situations.

"You've got to start thinking about protecting yourself, Ashley," stated Robert.

"I know that, Robert, but where do I start?"

"The way I see it is this: The main issue you're dealing with is between employer and employee. You're being positioned to do something unethical and/or illegal. You've got to know what your rights and obligations are, where you stand from a legal and an employment perspective. This situation could also very well lead to your leaving ZTE—either voluntarily or involuntarily—so you need to make sure your interests are protected in that regard too."

"Makes sense. So what do you suggest?"

"In my opinion, you need to think about going on the offensive. You need to talk to a plaintiff's attorney. In particular, someone who specializes in employment law."

"Do you have anyone in mind?"

Robert explained to me that he'd recently been through a contentious litigation in which the attorney for the other side was a guy named Harold Gregory. They'd been on opposite sides of the fence, but Robert thought highly enough of Harold to recommend him to me.

"This guy's sharp and he fights like hell for his clients. He's just the kind of lawyer you need."

The very next day, Robert met me at Harold's office. The office was in a beaten-down old two-story office building with a '60s-style façade. It didn't exactly fill me with confidence.

Harold Gregory was an older gentleman with a bit of a Wilford Brimley look and demeanor—a nonstop coffee drinker and a bit of a renegade, as most plaintiff's attorneys are. Sitting in his large conference room, he listened to me tell my whole story, which took over two hours, then leaned back thoughtfully in his chair. "You were right to come to me," he said. "You definitely have employment issues. When an employer asks you to do something unethical or illegal, you're the one in the hot seat, whether you agree or refuse to go along. Things rarely go in a happy direction for the employee. So, yeah, you've got employment issues, and I'll be happy to represent you on those. But you've got a bigger issue here, Mr. Yablon."

"Oh?"

"You've got potential criminal issues."

I felt instantly dizzy. It wasn't that I hadn't thought about the potential criminal implications of this case. After all, what ZTE was doing was massively illegal. But it's one thing to think about criminality, it's another to hear an attorney say the word aloud.

"So what do I do?" I pleaded.

"I want you to talk to a friend of mine," Harold said. "His name is Ted Masters and he's one of the best in the business."

The very next day I was back in Harold's office again, where I met Ted Masters. Ted is a very successful criminal attorney in Dallas. Like Harold, he is an older gentleman. White haired, tall, and mustachioed, Ted wears expensive suits with cowboy boots and a wide-brimmed hat. He also wears a hearing aid that he keeps well-hidden. Ted speaks in a whinnying Texas drawl that I'm sure has endeared him to many a Dallas jury. Ted has that "it" factor. You instantly like him, and you instantly feel better that he's on your side.

I went through my whole story again with Ted, then asked him if he thought I had criminal issues.

"Hell yeah, you've got criminal issues, Mr. Yablon. I'll put it even more bluntly: you have potential criminal liability here."

"Liability? What the hell? I haven't done anything wrong. I just learned about this whole scheme a few days ago. My concern is that if I go to the authorities, I'll be breaking attorney-client privilege. And that's potential liability for me too."

"Attorney-client privilege is an issue to consider, no doubt. But you're forgetting about a little something called the Crime-Fraud exception."

I waited for him to elaborate.

"The way I like to explain it is this: If you come into my office and say, 'Ted, I murdered my business partner last week,' I'm not allowed to say anything to the cops. Attorney-client privilege kicks in. But if you come into my office and say, 'Ted, I'm planning to murder my business partner next week,' that's a different story. I now have a duty to inform the authorities. As an attorney I'm not allowed to participate—either actively or passively—in the furtherance of a crime. I suggest you may be in the same boat here."

"Are you saying I have a duty to report my own client—or in this case, my employer?"

"I'm saying that's the issue you want to be thinking about. If you contribute to the furtherance of a crime by knowingly allowing it to go on, then, yeah, you've got some Texas-sized liability. And remember, we're not talking about shoplifting a stick of gum here. We're talking about crimes against the US government. We're talking about lying to a House Intelligence Committee. We're talking about falsifying a response to a government subpoena. You said yourself your bosses want to make you the scapegoat. That means they're going to do their damnedest to leave you holding the bag. But even if they don't—even if the worst thing you do is to know about their actions and fail to report them—you could still be liable."

"And what would my crime be, exactly?"

"How does treason sound?"

CHAPTER 11

The Walls Close In

I left Ted's office with my head spinning. *Criminal liability? Treason?* The very words created a lump in my stomach that made me want to throw up.

I climbed into my car, fumbled the keys out of my pocket, and loosely inserted them in the ignition—but I was too spent to even turn on the car. Each time I moved my hand to start the engine, another question would pop into my head: *How did I, of all people, end up in this position?* I'd shake my head in disbelief, unable to process my thoughts, then reach out to the car keys, and another question would come: *Why did I have to see that stupid contract?*

And another: *Why didn't I just look the other way?*

And another: *Why is this happening to me? Why? Why? Why?*

This went on for twenty minutes. When I finally did drive off, I soon realized the hows and whys were accomplishing nothing except to raise my blood pressure. I remembered what my dad used to say whenever we got ourselves into a predicament. It didn't matter how or why the situation had unfolded; the only thing that mattered was what you were going to do about it. Interestingly, this was the same advice I had given Meghan about the Reuters article—right before she told me how ZTE could no longer hide anything.

Suddenly I found myself pulling into the garage at home, no memory of how I'd gotten there.

I lay down on my bed and stared at the ceiling for what felt like hours, Gable sitting beside me the whole time like I was on suicide watch. Before talking to Ted, I hadn't quite realized—maybe because I didn't want to—how deeply screwed I might be just by keeping my mouth shut. I had been more focused on what would happen if I opened my mouth. Now it was becoming glaringly obvious that I was—pick your cliché—between a rock and a hard place; damned if I did, damned if I didn't; stuck between the devil and the deep blue sea.

In short, there was no clear way out of this mess, whether I spoke up or kept quiet.

Seeking counsel

For the next two weeks, I tried to soldier on at my job. I'd find myself reading an entire page of a contract, ready to turn to the next page yet realizing I had no idea what I had just read. The lights were technically on but no one was home. Even leisure moments—at work and at home—were spent wrestling with the question: *What the hell should I do?*

What was my move?

I sought counsel from every trusted friend, colleague, and family member I felt safe confiding in. Everyone was kind enough to lend me a sympathetic ear, but, just as with my own thoughts and feelings, their counsel was all over the map.

My mom was one of the first people I called. I didn't want to burden her with my troubles, but on the other hand, she was my mother and she needed to know what I was going through. Whether her advice was what I needed or not, it's always good to hear your mom say everything is going to be okay.

My mom, I love her dearly, is old school and really didn't see the full picture. She still believed in the *Mad Men*-era business fable in which you started out in the proverbial mailroom and worked your way up to CEO by putting your nose to the grindstone and working harder than everyone else. You were damn lucky to have a decent job. You worked for one company for a lifetime, and you didn't rock the boat. She was worried about my losing this job I had worked so hard to attain.

Our conversations would go something like:

Me: Mom, I just wish I knew what the hell to do.

Mom: What do your *employers* want you to do?

Me: I know what *they* want me to do. Keep my mouth shut.

Mom: Well, there you have it then. You work for them, not anyone else. They pay your salary—and it's your job to do what they hired you for. Your only responsibility in this matter is to . . .

Me: Just be quiet, keep my head down, and do as I'm told?

Mom: Ashley, I just don't want you to spoil this great opportunity. You have worked so hard for this. If you talk, you'll get a reputation as a troublemaker. No one likes a troublemaker. And you might never get a job like this again. I just want the best for you.

Me: Mom, we're talking billions of dollars in illegal international business.

Mom: I just want you to be safe, that's all. This is a very big company. You have a great title there—the title you worked so hard to achieve. I'm sure you can have a very nice, very secure career there if you just buckle down and do your job. Think about you and Donna. Why would you want to ruin that?

Me: Well, for one thing, because my company is selling spying equipment to enemies of our country.

Mom: I'm just worried about you; that's all. I want the best for you, son.

My mother is an extremely ethical person, don't get me wrong, but when she's worried, she hunkers down and goes into turtle mode. If I'm being truthful, her advice to keep my mouth shut was actually something I considered as possibly my safest option; maybe I should just pretend I never saw anything.

I kept hoping maybe something would happen that would make the decision for me. Or that maybe this would all just go away somehow, and I could happily continue my career as General Counsel for ZTE USA.

Right.

My brother Aaron was super supportive but had trouble seeing the international big picture. Aaron had always worked for large American companies, and, in keeping with Mom's advice, had been an ideal employee wherever he went. He was always the first person at the office in the morning, the last to leave, the guy who worked on his days off—and in that way he had worked his way up the corporate ladder. So the idea of an unethical act committed by the top leadership of a foreign company was unfathomable to him.

Me: Mom just thinks I should put a set of blinders on and play the "see no evil, hear no evil" card.

Aaron: Is she wrong about that? I mean, Ashley, you've got a great, great job. Giant telecom company, top of the industry. Why would you want to blow that up?

Me: I don't want to blow it up. I wish none of this had ever happened.

Aaron: So maybe it never did. Nudge, nudge.

Me: I can't just pretend I don't know anything and soldier on like a happy idiot.

Aaron: So quit then.

Me: And do what?

Aaron: Get another job. You must have a pretty amazing résumé by this point.

Me: I'm not some Wendy's employee who got in an argument with his boss. I can't just quit my job and walk across the street to work at McDonald's. It's a hell of a lot more complicated than that. I'm up to my ears in something really messy here.

Aaron: Not if you just quit.

Sigh. His intentions were good. And yes, to quit and find another position was indeed an option. Though not a particularly viable one, in my view.

My friend Robert, of course, was a great sounding board. He played the role of mediator and listener. He was one of my best resources too. He didn't try to minimize the potential trouble I was in; but he also tried to keep me calm and reassure me that I would get through this.

Me: I'm feeling so trapped, Robert. I feel like whatever choice I make, my life as I know it will be over.

Robert: I know it feels that way now, but this too shall pass. Are you any closer to making a decision?

Me: Not really. Ted Masters is telling me I've got criminal expo-sure and the noose is getting tighter by the minute. My family is telling to keep my mouth shut and be a good soldier or just quit and find something else. Meanwhile, Donna and I swing back and forth every five minutes. Stay/quit? Keep quiet/talk? We've got a mortgage to pay, bills, student loans, and a whole life plan that hinges on me keeping this job.

Robert: I think it would be wise to get a second opinion on the whole attorney-client privilege question. And a third opinion. Maybe a fourth.

Me: I can't afford any more lawyers, Robert! But you're right. I can't figure this out on my own, and maybe Ted is wrong. Sometimes when I look at the situation, it seems crystal clear: the contract was executed in the past, so my knowledge of it should be covered by attorney-client privilege. Therefore, I have a duty to keep my mouth shut.

Robert: On the other hand . . .

Me: On the other hand . . .

Robert: The crime is still going on. And you know about it.

Yes, damn it. The crime was going on. And I knew it. This was the whole furtherance of a crime/duty to report issue Ted had mentioned. And what an ongoing crime this was. Against our federal government, no less.

I had another lawyer friend who was also a great sounding board, though perhaps a little less diplomatic in his advice. Chad worked for a Chinese multinational company himself. He knew the ins and outs of Chinese legal thinking. He framed his advice something like:

Chad: These people are fucking crazy, Ashley! Fucking crazy! You've got to do whatever you can to protect yourself.

Me: What would you do if you were in my shoes?

Chad: Get the hell out of there! Lawyer up. Go on the defensive. Cut a deal with the feds. Assume the worst from ZTE. I work with these people, Ashley, I know the way they think and operate.

Me: Right. So what do you think their response will be if I rat on them?

Chad: These people are crazy, totally fucking bat-shit crazy. Protect yourself, whatever you do. Protect yourself!

Soothing words indeed. Donna and I, meantime, were engaged in endless mental wrangling along similar lines:

Donna: I just want you to be safe.

Me: I want us both to be safe. But what does safe" even mean anymore? When I think about protecting us—our home, our life together, our financial safety—it's clear to me that I have to keep quiet. Keep the job at ZTE and try to get a new job as soon as possible. No one will hire me if I become an international whistleblower, that's for sure.

Donna: But if you keep quiet, you might end up in jail. And we'll lose everything anyway.

Me: But if I try to play it safe by going the other way . . .

Donna: . . . by talking to the government . . .

Me: Then I'm going to make an enemy of an extremely powerful group of ZTE guys with ties to the Chinese government and who knows what other organizations. As Chad is fond of pointing out, these people are fucking crazy. And I don't want us to end up dead.

Donna: Fuck, Ashley.

Succinctly put.

Ted Masters—to whom I was talking every other day—was the sober voice of reality, speaking the one message I really didn't want to hear.

Ted Masters: You've got to go to the FBI, Ashley. Time is running out. I can try to get you immunity, but you're going to have to act soon. The longer you wait, the more that option goes out the window.

Every time he'd utter those letters, F-B-I, my heart would do a cartwheel in my chest. FBI? Was that really my best option? Talking to the freaking FBI? Once I did that, I would be crossing the Rubicon. There'd be no going back.

A big part of me knew Ted was right, but I couldn't make peace with the idea. I was like a person who needs to lose weight reluctantly seeking a nutritionist's advice. They keep telling you that you need to go to the gym and change your eating habits; you keep coming up with excuses as to why that's really not possible right now. The idea of talking to the FBI was a truth I couldn't accept, and I would do anything to justify why it wasn't an option.

The reality was I was making excuses because I was scared to death.

Ted Masters: You have issues here, Ashley, serious issues. And each day you wait before taking action, the shit gets deeper. It's been about ten days since you saw that contract in China. Okay, fine. That's still a reasonable time period for coming forward. But the longer you wait, the more your potential liability increases. This crime is ongoing. The longer you wait—two weeks, three weeks, four weeks—the more this shifts from you keeping silent to you doing something far graver.

Me: Meaning what?

Ted Masters: Meaning you'll be actively involved. With every email or phone call you receive now in which you knowingly play along with the status quo, you become, more and more, an active participant in the furtherance of a crime. It's no longer their crime; it's your crime as well.

Me: Shit, Ted. Isn't there anything we can do short of going to the FBI?

Ted Masters: We can try approaching the Texas state bar, ask them for an official opinion on what your ethical obligations are.

Me: Good, yes, let's do that.

Ted Masters: But it might take three or four months to get a ruling. We don't have that kind of time. No, the situation you're in here is one that I would best describe as—not to put too fine a point on it—a ticking time bomb.

Ticking time bomb indeed. That was exactly what my life felt like. The knot of anxiety I was carrying in my chest was threatening to tear my ribcage apart. And I couldn't find any relief from it, day or night. I couldn't eat. Sleep had become a joke, and my lack of rest was catching up with me. My mind was starting to play tricks on me, and I was becoming fatigued and forgetful. Walking Gable, I'd be lost in thought, failing to notice the oncoming traffic swerving out of the way to miss us.

There was no clear answer here. No definitive right and wrong. Tons of gray hues—each with its own set of problems.

Meanwhile at the office . . .

The situation at work was only exacerbating my stress. I was trying to do my job—managing the legal issues of the company, dealing with contracts and litigation matters around the country, as well as employment issues—while also keeping up with the HPSCI investigation and the onslaught of emails I was receiving from the Department of Commerce and the US component-part manufacturers. I found myself looking over my shoulder every minute, jumping out of my skin every time someone touched my arm. I was worried sick about receiving a phone call or an email that would call for me to take any sort of action—direct or indirect—in relation to that Iran contract. All I could hear in my head were Ted's warnings: any move I might make related to that issue could be construed as my being a willful participant in the furtherance of a crime. And if I refused to take such action, I would instantly bring matters to a head with my bosses and force a confrontation. So every time my phone rang, my blood pressure spiked. To this day I still jump every time I hear a phone ring or a knock on a door.

We still didn't know how things were going to play out regarding the subpoena from the Commerce department. When I'd been in China, as you recall, I'd been told by Guong One that ZTE China intended to fully comply with the Department of Commerce's subpoena, only to be later informed by Meghan that that was a huge lie. Since returning to the States, I'd been hearing conflicting rumors as to whether the company was really going to cooperate or not.

DD&M had asked Commerce for more time to respond to the subpoena, and Commerce, in return, had given us until April 20 to either produce the wanted documents or provide a letter officially stating that we would provide them by a specific date. As of April 19, nothing had yet been done about this, so that was causing acute anxiety for me.

Finally, at the eleventh hour, DD&M contacted the Commerce department, stating we would provide the letter promising compliance with the subpoena.

Hallelujah. This was a massive relief. It meant maybe I was going to dodge this bullet. Maybe everything was going to resolve itself without implicating me at all. You see, if ZTE really did plan to turn over those subpoenaed documents, then the Department of Commerce would soon learn for itself everything I had learned in Preston's darkened room. They would see the contract directly. They would see the part that read, "How We Will Get Around US Export Laws," and they would learn all about ZTE's use of shell companies to skirt US embargoes. They would learn all this stuff without my having to tell them. And if so, that meant I would likely be off the hook. After all, I wasn't implicated in the contract itself. I'd had nothing whatsoever to do with the drafting of that document, and I had never held a copy of it in my hand or even been permitted to download it onto my computer. My only crime had been laying eyes on it and acquiring knowledge about it. But now maybe that was all moot. ZTE's nefarious deeds would be exposed directly. That would be bad for the company, but very, very good for me.

That weekend, I was feeling something approaching optimism for the first time since I'd left Preston's office in Shenzhen. I remember going out with friends for dinner on Sunday, April 22, and having some hopeful phone conversations with my family. I think I even laughed a few times. I was upset that the company might suffer major repercussions and that

I might have to start looking for another job, but that was a hell of a lot better than a stint in a federal corrections institution.

Coming to a head

My cautious optimism was short-lived. On Monday, April 23, I entered the office to find a small crowd of employees huddled around a computer. I walked up to the group and asked the sole American what was going on.

"If this isn't targeted directly at us, then I don't know what is!" he exclaimed, pointing accusingly at the screen as his Chinese counterparts looked on in confusion. President Obama had just signed Executive Order 13606. And as I read it, I saw that it was indeed an order directly targeted at ZTE. The order blocked the property, and suspended entry into the US, of any entity that operated, sold, or otherwise provided technology to Iran or Syria—technology that could be used to conduct computer or network monitoring and thus enable human rights abuses by the governments of Iran or Syria.

The order was a clear message from the US to China, to ZTE, and to me: you are front and center on the US's radar screen. We are coming after you, and you are a priority target.

Holy shit, when the President of the United States throws the gauntlet directly at you, that's not good.

Things continued to get weirder two days later.

On April 25, just a couple weeks after the China trip, Matthew came into my office. I had been overhearing him speaking in Chinese across the hall, so I knew he had just gotten off the phone with China. "Hey, Ashley," he said to me in a casual voice, leaning against my doorframe, "Can you please delete all those emails involving that conversation with Preston, anything related to exporting to Iran or any other country? And shred any paper copies?"

I couldn't believe my ears. Normally, I would have chalked up such a request to Matthew's naiveté, but I knew the request hadn't come from him. It had come from China.

"No," I replied to him, lacing my voice with incredulity, "of course I can't."

He paused, looking as if he was going to ask for further clarification, then just nodded thoughtfully and said, "Okay." He walked back into his office, and a moment later I heard him speaking on the phone in Chinese again.

Seriously?

And then the other shoe dropped. Later that day I started hearing vague rumors about the subpoenaed materials. No one seemed to want to talk to me directly about the issue, though. Again I was getting that uncomfortable sense that I was being excluded from communications. More than once I walked into a room where Meghan or a ZTE executive was talking, and everyone would go instantly silent.

Finally, I buttonholed Meghan in a hallway and asked her what the hell was going on. Her tone these days when giving me "magic" was no longer that of a confidant sharing intel to help an ally in the trenches. Now, "magic" was shared only sparingly—and mostly through my coaxing. My friendship with Meghan appeared to be melting away. Was she concerned I might talk? Or did she know something and was trying to protect me by freezing me out? I didn't know, but I could tell she was hesitant to say anything. Finally, she spilled the beans.

"Magic. We've received word from China that neither ZTE China nor ZTE USA will be complying with the subpoena."

"What?" I said, my head jerking in disbelief. "Not complying? We've already promised the Commerce Department, in writing, that we will comply! We've promised to send them everything!"

"Change of plans. Our new position is that it is against the laws of China to comply with such a request, and so we must honor Chinese law. Magic."

God damn it. Instantly I was back on the hot seat again. I was once again the only means by which the US government would be able to learn the inside details of ZTE's dirty dealings with Iran. I stormed to my office and got Jeff Pennington on the phone.

"Jeff! What's this I'm hearing about us not complying with the subpoena? And why is nobody telling me about this?"

"Don't be paranoid, Ashley. We were going to tell you. We just wanted to get our ducks in a row first."

"Why the sudden change of heart about the subpoena? You do realize the position this puts me in, given what I told you I saw when I was in Shenzhen?"

Pennington paused meaningfully. "We know what you think you saw, Ashley, but we also know you don't have anything in writing. There are some of us who think it's possible you were mistaken about what you believe you were seeing over there. It's possible your memories have become . . . distorted."

"What are you talking about, Jeff?"

"Well, for one thing, I've been in talks with China and they assure me the component-parts in question were never actually shipped to Iran. They assure me those parts are, in fact, sitting in a warehouse in China."

"But that's not true."

"My client says it is."

"God damn it, Jeff, what *the hell* are you talking about? You know damn well those parts were shipped. Mr. Guong asked me, right to my face, what kind of cover story we should come up with to explain the shipment away. Your guy Nueve was standing right there. Guong was talking about flying a team to Iran to replace all the US parts or scratch out all the serial numbers. Those materials were shipped to Iran and we all know it!"

"Not everyone remembers it that way."

"No, no, no. Don't you dare do this, Jeff. Don't you dare pull this gaslighting crap on me. Don't you dare start trying to make me out to be the crazy one here."

He paused. "I almost said, 'If the shoe fits,' but that would be unkind."

"Fuck you, Jeff. So is that the new party line amongst the thousand-dollar-an-hour crowd? Ashley's become a little unstable. We can't trust him anymore. Better not tell him too much. Is that why I'm being systematically cut out of the communication loop again?"

"No one's cutting you out of anything. Listen to how you sound."

"Stop it! I'm being ghosted by the whole team right now and you damn well know it. At least have the decency to admit that's what's going on and stop lying to me."

There was a long silence at the other end of the line.

"Fair enough, Ashley, I suppose you deserve some honesty. A late-night phone meeting was held several days ago. Some of the DD&M team members were on the line, as were Meghan, the two Mr. Guongs, and several members of the Chinese legal team. At that time, it was decided that you were no longer to have any involvement in matters related to the subpoena and the Department of Commerce."

"What? Why? Because I'm crazy and can't be trusted? No, I'll tell you exactly why. Because I'm the only one who doesn't feel comfortable going along with the Big Lie. Because ZTE is worried it can't count on me to be its scapegoat any longer, so they're freezing me out. I'm going to talk to Meghan. She'll confirm that that's exactly what's going on here. Meghan's the only person I can trust in this fucking bunch."

"Meghan?" said Pennington. "She was the one leading the charge to freeze you out."

I hung up the phone and felt the bottom drop out of my world.

Numb and speechless, I wandered out of my office and down the hall. I pushed the elevator button about twenty times, and when the car finally arrived—after what seemed like a week—I hit the *G button and headed the for the ground floor. I stepped out into the hot parking lot and made a beeline for my car. Sitting in my parked car with the air-conditioning on, I pulled out my phone and found Ted Masters on my speed-dial screen.

"Ted, it's Ashley," I said, hushing my voice for no reason. I proceeded to tell him the latest bad news on the subpoena front and about the ghosting job I was getting from the rest of the legal team.

"You know what your next step has to be, my friend," said Ted. "You've known it all along."

"But there's so much at stake here, Ted! I don't know how I can do it."

I'll never forget his next words. They changed my life in every way.

"Ashley," said Ted, adding a pregnant pause, "I don't know how you can't."

CHAPTER 12

Rubicon Crossed

I don't know how to describe it, but when Ted Masters said those words, it was as if a light switch flipped on in my head. All the mental anguish of the previous two weeks receded, and clarity emerged. No longer was I clouded with doubts or second-guessing. I knew what I had to do. I knew my decision would change everything in my life—my job, my future, my reputation, my sense of safety in the world—but I also knew there was no other choice.

"Okay, I'll talk. Let's do this," I told Ted.

An hour later, Ted called me back on my cellphone. I quickly shut my office door before accepting the call so as not to have "cubicle-land" overhear me.

"Tomorrow, my office, 1:30," he said. Hearing those words felt like getting a call from your doctor telling you your test results are in and you need to come to the office right away.

I was terrified. But ready to take the dreaded step.

Bean spilling

As I have mentioned, Ted is a criminal attorney, so his office takes certain precautions that a patent attorney's does not. You don't just stroll in unannounced with a box of donuts hoping to chat. Rather, when you open the door from the hallway, you find yourself in a small waiting area with a locked inner door, a doorbell, and a speaker (probably a hidden camera too). You press the button, say your name, and then you wait to be let in. Or not.

Going through that little ritual in the early afternoon of May 2, 2012, felt a tad more ominous than it normally might have. *Seriously, I'm at my criminal attorney's office, about to meet with the FBI?*

My only other face-to-face meeting with Ted had occurred at Harold Gregory's office, and all our subsequent discussions had been over the phone, so I was new to his office protocol. After reading the sign and ringing the bell, I paced nervously for a minute or two, waiting for something to happen. Ted's receptionist finally opened the door and greeted me with a smile. "Good morning, Mr. Yablon, please follow me." She led me past a small group of cubicles and past Ted's office—filled with eclectic antique furniture and photos of Ted presenting his cases to Dallas criminal juries—to the conference room, which featured a floor-to-ceiling window looking out on the North Central Expressway (the same freeway that bordered the ZTE USA office several miles North).

As the receptionist showed me into the conference room, Ted quickly stood up from the table and approached me from my left. He took hold of my arm reassuringly as my eyes became temporarily blinded by the brightness of the afternoon sun filling the room. I could vaguely make out two figures silhouetted by the huge window. As Ted led me around the table, my eyes began to adjust to the light, and I noticed the two individuals rising in unison.

"Ashley, I'd like you to meet agents Zander Caldwell and Jessica Baxter from the Dallas office of the Federal Bureau of Investigation." Maybe Ted's use of the agency's formal title was supposed to sound less intimidating than the stark "FBI." It didn't.

The agents sat, and I was invited to do likewise. As I parked myself in the antique Chippendale chair, my eyes now fully adjusted to the brightness of the afternoon sun, I was able to see the agents more clearly. Zander Caldwell was around my age, maybe a couple of years older—tall-ish, slender of build, looking like a former high school athlete who'd kept himself in pretty good shape. "Agent Caldwell," said Ted, "specializes in counterintelligence and counterespionage investigations, including investigations of federal export laws."

"Agent Baxter," Ted continued in his Texas drawl, "specializes in those same areas." Baxter nodded. She was also close to my age, perhaps a bit younger, with dark mid-length hair that had a bit of a wave. She had a kind, inviting look and persona—certainly not what I'd envisioned for an FBI agent. "We have a lot to accomplish today, so I suggest we dig right in."

"Pleasure to meet you both," I fibbed with a light smile.

Now, as an attorney, I have had my share of nervous moments in front of a jury or a judge, where you do your best to try to maintain a cool, calm, and collected demeanor. You're like a swan: composed and elegant on top of the water but a paddling, whirling dervish below. I've played that role before.

But this? No tense cross-examination of a witness or sweating out a jury's decision could have prepared me for this.

My God, I'm sitting down with the fucking FBI. Is this really happening? was all I could think. The urge to panic welled up inside me, but I managed to tamp it down with a couple of deep breaths.

I noticed a small digital recorder on the table with its red light already on. Caldwell saw my eyes land on it.

"We'll be recording everything today, Mr. Yablon, if that's okay." I'm thinking: *Right, and if I say no?*

"Later on, we will be using any information you give us to create an affidavit, which we may use to apply for a warrant to conduct some physical investigations. Do you understand this, and are you in full agreement?"

"I am."

"Great. Then let's get started."

Caldwell detailed his background a bit more, explaining that he was experienced in various methods of investigation including electronic and physical surveillance, the executing of search warrants, and the handling of Confidential Human Sources (CHS), and that he had personally directed investigations related to matters of national security. The time he took to detail his background allowed my nerves to calm a bit more.

"That's my story," he said, "now let's hear yours."

"Where do I start?"

"The beginning is usually a good place."

I began to talk, guardedly at first, but pretty soon the floodgates opened, just as they had with Ted and Harold Gregory before. It took all afternoon to get through the bulk of my story—the agents would stop me here and there, asking for clarification and elaboration on certain points or to check the recorder. As the afternoon progressed, my nervousness dissipated and the unburdening became easier; it actually felt pretty good. I began to see this session as something of a catharsis after those weeks of strain and fear. Maybe I had made the right choice after all.

Every hour or so, one of Ted's staffers would bring me a fresh bottle of water. My throat was growing progressively drier the more animated I became in my description of the events.

Toward the end of our conversation, my new federal friends asked if I had in my possession a company-issued laptop. They wanted to "borrow" it. I assumed they were keen to see what my e-mails, documents, and other records had to say relative to the Iran contract and ZTE's efforts to evade US embargo laws. But since I had tried to leave the office inconspicuously, I had not brought my laptop with me.

"Why do you need to take physical possession of my computer?" I asked.

"So we can make a preservation copy."

"What's that?"

"It's an official copy of the contents of your computer as of a fixed date," said Baxter. "At some future time we may need to officially seize your computer, along with other work materials. So in case anything were to disappear from, or be changed on, your hard drive, we would have a dated copy on record that can't be altered."

"Okay, I can bring my laptop when we reconvene in a couple days."
We had already agreed to meet a second time. "Will that work?"

"Sure," said Caldwell. "Let's aim for the afternoon of May 7."

So on May 7, I came back to Ted's office and completed the last part of my
sit-down with Agents Caldwell and Baxter. When we finished, they asked
about my laptop.

"I brought it; it's in my car."

"We'll walk you down."

On the way down in the elevator, the agents explained to me, again,
that everything I had told them would be captured in a written affidavit,
but that this affidavit would be placed "under seal" and would be absolutely confidential. No one would ever see it. And if a judge were to review
the affidavit during a closed hearing to grant a warrant to raid ZTE or
the like, that hearing would also be under seal. It would never be public
record. No one at ZTE would ever know I was the one who had spilled the
beans on their illegal conduct.

I was still afraid, of course, that ZTE would figure it out anyway. Trust
had already become a major issue between me and my employer.

Caldwell, Baxter, and I headed into the parking garage, where we
huddled around the trunk of my car. I found myself looking suspiciously
down the rows of cars and doing a nervous three-sixty. Silly me; nothing
bad ever happens in a parking garage, right?

I dug through my briefcase, grabbed the computer, and proffered it
to Agent Baxter.

As we both held onto the laptop's edges, I explained to the agents, "I
use this computer every day. It has all my work files on it. I need to bring
it to the office with me tomorrow so I can do a normal day's work."

"Don't worry," said Agent Baxter. "We'll have our forensics team
work on it overnight, and they'll be done with it by morning."

Here's the interesting thing: never once did they ask me for my computer logon name or my password. Child's play for the FBI, I guess.

Agent Baxter set the laptop down and took out her notebook to use as a writing surface. Producing a three-ply white/yellow/pink duplicate form, she flipped over my laptop and wrote its serial number on her form—which I now noticed was a receipt. She handed me a pen and the receipt to sign, then peeled off the top white copy and handed it to me.

That's right: I had just been given a receipt for "one laptop computer," signed by the FBI, as if we'd just done a Craigslist transaction. Again, can't make this stuff up.

"So how will I get it back?"

"We'll be in touch, Mr. Yablon."

"When? Where? How?"

"We'll be in touch. You did good today."

That was the best they could give me? I guess "did good" was code for risking your life and committing career suicide.

It was about 5:30 p.m. I drove back to the office before going home, to see if there was any urgent business I needed to attend to—but also just to assure myself that my company security badge still worked; that no one had found out about my secret FBI meeting or quietly had me terminated.

Everything seemed normal as I entered the office floor—everyone who passed me gave me an innocent nod or smile. Great. No one was on to me. Yet.

I went into my office and sank down in my desk chair, already starting to feel like a stranger in my own workspace. Somehow I knew my meeting with the FBI had been the first domino to fall in the official collapse of my career as General Counsel for ZTE USA. I knew it.

I went to boot up my computer. *Oh wait, it's in the hands of the FBI.* Geez, I had driven all the way to the office for nothing? The freak-out of meeting with the FBI had scrambled my brain.

Using my cellphone, I downloaded my email. One message jumped out at me and made my heart sink. It was from Agent Fulmer of the Department of Commerce. He wanted to meet with me. Personally. ASAP.

A perfect ending to a perfect day.

Where's the movie camera?

I awoke the next morning in a surreal state of mind. Had I actually gone to the FBI yesterday and ratted out my company, or had that been a dream? Sadly, I recalled it was the former.

"So how does it feel to be a multibillion-dollar whistleblower?" Donna asked, sipping her Sugar-Free Red Bull, trying valiantly to put a light spin on things.

"Not bad," I said, then gestured toward the window and added, "You'll forgive me if I keep the shades drawn this morning, though. Don't want to make the sniper's job any easier than it has to be."

We laughed our asses off. Sometimes that's all you can do.

"I am so sorry I put you in this position," I said to Donna. "This was not part of The Plan back when we were at Loyola. Where the hell did I go wrong?"

"You didn't go wrong. You got screwed. And who knows, maybe this is part of The Plan."

"Maybe," I owned. I doubted it, though. It was hard to believe there was any silver lining or cosmic purpose in this thing.

"You going to work today?" she asked me.

"I don't know. That was the plan, but the fucking FBI has my work computer."

A little while later, I was taking Gable for his morning walk when I saw a call come in on my cell phone from "Blocked." *Who could that be?* I wondered. *Some bill collector somewhere? Think I'll pass.* After the call repeated itself a couple more times, though, it finally dawned on me who was calling. I noticed they'd left a voicemail already, and I tapped Play. "Hello, Ashley. This is Agent Baxter. Agent Caldwell and I are at the Starbucks in Inwood Village right next to your house."

I trotted home with Gable, threw on the Uniform, and hurried around the corner to the Starbucks. Our local Starbucks was the type that had outdoor seating on a sidewalk patio. As I approached the shop, I saw Baxter and Caldwell sitting at one of the wrought-iron tables under a green umbrella. They waved at me, smiling, as if we were friends or neighbors. Got to play for the crowd, after all.

As I sat down to join them, Caldwell said, "We got you a coffee. Black, no sugar or cream, right?"

He handed me the cup and I took a sip. I had never had coffee with these guys before. "How did you know that?"

Caldwell gave me a little smile that said to me, *Who do you think you're dealing with here?* Message received.

The coffee was a gentle warning delivered in the form of a thoughtful gesture.

We made chit-chat for a few minutes—the weather, baseball, our mutual plans for the weekend. The agents were trying to make our meeting look like a social get-together, though I do think they were also making a genuine effort to be neighborly.

After a few minutes, Baxter reached down beside her chair and grabbed, of all things, a large Happy Birthday gift bag, complete with colored ribbons and puffy tissue paper.

"This is for you," she said, deadpanning.

I gave Baxter a curious smile as I accepted the gift across the table. It felt awfully heavy for a tie or a box of chocolates. That was when I peeked inside and, through the ruffled tissue, spotted my laptop. So this was how they had chosen to return my computer to me? Stop me if I've said this before, but you can't make this shit up.

Reflexively, I looked around, up and down the street. Where was the camera crew stationed? Because this felt like a scene right out of a movie.

I said my farewells and headed off with my Happy Birthday bag.

The screws tighten again

I drove straight to ZTE with my coffee and my laptop. Again I feared, as I ascended in the elevator and used my electronic badge to enter the company offices, that somehow I had been "found out" already. I half expected to be descended upon by burly Chinese security guys who would carry me out the door and throw me in the back of a GMC Yukon.

No such thing happened.

I passed, unnoticed, through the huge cubicle farm that occupied much of the floor and headed to my office. Sitting at my desk, I pulled up the email from Agent Fulmer of the Commerce Department—the one stating that he wanted to meet with me personally. I think I wrote him some kind of noncommittal response about setting up an appointment shortly, as soon as I could blah, blah.

I leaned back in my chair and let out a sigh. I was feeling squeezed again. The Commerce Department was continuing to target me for the subpoena materials. *Why?* I mused. Part of the reason, of course, was that I was General Counsel, but I suspected the larger part was because I was the only ZTE executive who was not a Chinese national. I think they figured that I, as a patriotic, English-speaking, American citizen, represented their best shot at convincing the company to play ball. In their eyes, I represented the company's vulnerable spot—the crack in the armor they needed to get their documents. But I had already cooperated with the FBI—shouldn't that be enough to get the Commerce Department off my back?

From the China side, I was still hearing that ZTE did not intend to comply with the subpoena. So I found myself stuck in the middle—hoping to convince the ZTE execs to change their minds, while also trying to buy time with the Commerce people. In my mind, I still had a chance to avoid becoming the identified whistleblower in this mess—if only the company would surrender those damn documents.

Things went on like this for several more days. The House Intelligence investigation continued to unfold, DD&M continued to bill the living crap out of the company, and I continued to do my job and to dance on the electric fence between ZTE China and the Commerce Department.

Follow-up emails poured in from Fulmer, and they were making me nervous. Why was he so insistent on meeting with me personally? One day in mid-May, I sent an email to Guong One, Guong Two, and Meghan, telling them I felt the time had come for me to retain my own attorney. I also told them the company would be responsible to cover the costs of that attorney. In response I received only e-hemming and e-hawing— noncommittal avoidance of the issue.

Pressure from the US government continued to mount. At one point in late May or early June, Katherine Van Bergen of the HPSCI sent an email to the company expressing a desire to meet with the most senior person in ZTE USA. Meghan and the DD&M team brainstormed about who the best person might be and came up with an ideal candidate: me.

When I learned about this, I nearly blew a valve. I stormed across the hall to Meghan's office and said, "What's this about me meeting with the House Committee? Is everyone still entertaining some fantasy that I'm going to stand up and lie to the government to cover ZTE's ass? I'm not the senior executive here; Xi is. I'm not the expert on the House issues; DD&M is. I wasn't even present for the House meetings in Shenzhen. You guys systematically exclude me from conversations and freeze me out of the information loop, then you expect me to be the fall guy who takes the heat? No, thank you!"

Meghan said nothing. I turned and walked out.

It was finally decided that Xi would be the most appropriate person to meet with the House committee.

Meanwhile, Fulmer from Commerce was continuing to insist on an in-person meeting with me.

I couldn't avoid him any longer. I got with him on the phone, and he told me, in effect, "The Commerce Department is getting tired of ZTE's song and dance. We issued your company a subpoena months ago, and now we're pulling teeth to get you to respond to it. What the hell's going on?"

"It's complicated, as you know," I replied. "Not only is China not technically required to comply with a US subpoena, but Chinese law specifically forbids such compliance."

"Right. Listen, Mr. Yablon, here's the bottom line. We believe you can talk some sense into your Chinese bosses. We believe you can convince them that it will be bad for ZTE USA if the company continues to play hardball. And what's bad for ZTE USA is ultimately bad for China. Perhaps very bad."

"I think you assume I have more leverage with China than I really do."

"Well, I'll tell you this, Mr. Yablon, we're fed up with the game-playing. If we can't get ZTE to cough up the materials we need to see, then we are coming after you personally. Do you understand what I'm saying?"

Crap. I understood perfectly. I was now in the crosshairs. Not just as General Counsel of the company, but as Ashley Yablon, private citizen. I thought back to my time at McAfee and how the former GC there, Kent Roberts, had been held personally accountable for some of the actions he had taken in his role at the company. A chill ran through me. I didn't think I had done anything to create personal liability for myself, but I was no expert in such matters. It was time to hire someone who was. I needed a lawyer. Yes, another lawyer.

And this time, the company was going to pay for my attorney, damn it.

What ensued next was a series of emails and phone calls with my US bosses, my Chinese bosses, and the Chinese legal team. I explained to them that I needed an attorney to represent me individually. They responded that I should just use DD&M.

I pointed out the obvious conflict of interest in this idea. DD&M was currently doing millions of dollars' worth of legal work for the company. How could I expect them to fairly represent me as an individual, especially if my interests clashed with the company's interests in some way? Again my bosses failed to see—or pretended to fail to see—any inherent conflict.

I produced documents showing them that I was entitled to my own representation under the company's bylaws and D&O (Director and Officers) policy. (If you're not familiar with D&O insurance, it's a policy that protects a company from claims that can arise due to the actions, inactions, and decisions of members of the company's management team. It's management insurance, essentially. And I was part of management.)

No matter how I pleaded my case to the ZTE brass, though, the answer I received was: *No, we're not going to pay for your attorney.*

To which my response was: *Then we have a problem here.*

A new line had been drawn in the sand.

I decided my only recourse was to file directly for coverage through the company's D&O policy. I did the paperwork and sent the claim in to Chubb, the issuer of the policy. The Chubb people promised to give me a speedy response, but I knew it would be at least a week or two before I heard back from them.

Toward the end of May, HPSCI met with ZTE again in China. Xi went to Shenzhen to represent ZTE USA. I stayed home. On June 12, Xi received a letter from HPSCI stating that the Committee had an additional twelve pages of questions it wanted answered and documents it wished to see. And it wanted everything in three weeks' time. *Good*, I thought, *the more pressure on the company, the better.* I knew the wheels were going to come off this bus soon, and I was still desperately hoping they would come off as a result of the Commerce probe and the HPSCI investigation, not as a result of my FBI chat.

On the same day Xi got his letter, I received a letter from Chubb denying my insurance claim.

Screw this, I thought. *I am NOT going to be scapegoated here. Not for something I didn't do. Not for refusing to participate in a cover-up.*

I needed to find a way to stop playing defense and go on offense.

CHAPTER 13

We Now Rejoin Our Story in Progress

The fear of being personally targeted by the Commerce Department was weighing heavily on my mind. Before Fulmer had even started threatening this action, I'd already been afraid I was being set up by ZTE to be its scapegoat. The company's refusal to help me obtain my own legal counsel had only doubled that feeling. What it told me was the company didn't have my back. I was afraid not only of being hung out to dry, but of being turned into a sacrificial lamb.

After Chubb refused my claim for an attorney, I went back to my informal "advisory board"—my lawyer friends—and asked them what I should do. It was Chad who gave me perhaps the shrewdest and most important piece of advice I received throughout this entire debacle.

A new tack

"You need to sue them," Chad told me one evening as we chatted on our cells on our drive home from work.

"Sue them?" The idea hadn't even crossed my mind. "Now? Why? For what?"

"For failure to provide you with legal representation. Now that the Department of Commerce is coming after you personally, not just in your role as GC, you're legally entitled to have ZTE pony up for your representation. That's a very specific area where you've got a clear case against them." Chad was an employment lawyer; I valued his opinion immensely.

"But I'm still working at ZTE," I said, "and I'm already on thin ice as it is."

"All the more reason to sue them now," replied Chad. "Let's face it, Ashley, you and I both know this thing isn't going to end well for you. That writing is already on the wall. You need to sue them now; give yourself some leverage with them as you exit. Force them to deal with you seriously. If you wait till after you leave the company and then try to go after them for damages, they'll have no motivation to settle with you. They'll just out-lawyer you. A lawsuit might be the only thing that protects your ass on the way out the door." Maybe he was right.

The idea of suing my current employer felt strange and uncomfortable. I definitely needed a second opinion. I ran the idea by a couple of other members of my advisory board.

"It's brilliant," said Robert, whose opinion I also valued immensely.

"I think it's your best move, frankly," echoed Ted, giving the idea a tip of his hat.

Okay then, it was settled. I would sue ZTE, even as I was still employed as its General Counsel. This was not going to land well with my employers, I knew. But then again, neither would the fact that I had gone to the FBI, a truth that seemed destined to come to light very soon. And let's face it, my stock with the company had already gone down precipitously due to my refusal to play along with my bosses' scheme to make me their designated liar. ZTE was probably looking for ways to get rid of me anyway. So how much worse could things really get?

A dumb question to ask, I realize.

I started hitting Google. I quickly learned that suing my employer was not going to be simple. Or cheap. For starters, ZTE USA, though headquartered in Texas, was incorporated in New Jersey, which meant I would need to conduct my suit in a state 1,500 miles away. Did I really want to take on that hassle and expense? I was already paying two employment lawyers and a criminal lawyer and was borrowing money from friends and family to do so. How could I possibly afford to take on another lawyer's fees?

But then again, if Chad was right about the lawsuit being my only real leverage against ZTE, how could I afford not to? Damn this whole mess.

I started researching law firms in New Jersey and came across one that, astonishingly enough, offered representation for "General Counsels or corporate executives whose employers fail to provide attorneys' fees for them" as one of its specialties. Such weird specificity could not be ignored. The firm was called Peters, Sullivan & Bain. I gave them a call.

The gentleman I spoke to, Andrew Silver, was a fairly young attorney—younger than me—but pretty sharp and knowledgeable. I told him my story, and he thought my case had merit, at least initially. "In order to move forward, I'm going to have to file a petition and serve your employers. We're going to require a check for five thousand dollars."

Sure, why not? What was another five grand?

Donna and I didn't have the money, of course. Well, not in full. I threw in whatever I could come up with and Donna needed to put in some of those extra hours she used to work when we were still in acquisition mode. But we felt like we had to play this thing out.

So just like that, ZTE was served with my papers. I had now officially become a plaintiff in a lawsuit against my powerful, deep-pocketed, Chinese government-connected employers. It felt pretty damn strange, as I knew it would. An article about the case came out in a well-known law periodical—a Texas GC suing his employer—and I'm sure a lot of people who weren't in my shoes derived some educational or cautionary benefit from it. I did not.

I was still on the hot seat and feeling pressure every day. I had taken a huge personal and professional gamble by going to the FBI and doing the right thing. But now the Department of Commerce was trying to use me as its lever to get ZTE China to cooperate, and threatening to indict me personally if it didn't get what it wanted. It was as if one hand of the government didn't know or care what the other hand was doing, and I was the one in the middle, getting squeezed. I expressed my vexation about this to Ted on more than one occasion.

Ted heard me. On one of the first days in June, he called me and said he had set up a meeting with some people I needed to talk to. He didn't tell me too much else; Ted tended to share information only on a need-to-know basis. And evidently I didn't need to know anything about that meeting till I got there. All Ted told me was when to show up and to not be late.

The "triangle" meeting

If I recall correctly, the date of the June meeting was exactly eight months to the day since I had started my job at ZTE. I remember musing about that as I rode the elevator up to Ted's office floor that day. Eight mere months since the high point of my professional life. Eight months since I had landed my dream job as General Counsel for ZTE, one of the largest telecommunication corporations in the world. Eight months since Donna and I had celebrated with cake and Veuve Clicquot. Eight months since she'd told me, "You've got the world by the tail now, Ashley."

Eight short months. And now where did I find myself? Walking into my criminal lawyer's office for a secret meeting with some people whose identity Ted wasn't sharing. Were these more heavy-hitting government agency folks? Whoever they were, the whole scenario felt insane.

I pushed the button in Ted's secure waiting area and announced myself. The receptionist let me in with her usual smile and led me to Ted's antique-filled office.

As I entered the room, I looked around. Four individuals were seated in a rough circle; two of them I knew, two I didn't. Ted invited me to sit in one of his Chippendale chairs and assumed a standing position behind me, placing a steady hand on my shoulder as if to say, *Don't worry, everything will be fine.* Which, of course, made me fairly certain everything wouldn't.

Ted directed my attention first to my left, where, sunken almost comically into a squishy tan leather sofa, sat the two FBI agents to whom I had officially blown the whistle on my huge Chinese multinational employer a few weeks earlier.

"You know agents Caldwell and Baxter," Ted said.

"Good to see you both again," I said. What I was actually thinking was, *This whole nightmare has turned into a Russian novel; it started off bad and has only gotten worse.*

Ted then gestured toward a round man in a gray suit sitting uncomfortably across the coffee table from me in another Chippendale chair. "This is attorney Stevens from the US Department of Commerce's International Trade Division." Stevens raised his saucer and tea cup slightly, nodding in my general vicinity but not going so far as to make eye contact.

To my right, in a little club chair, sat another gentleman, dressed in a blue suit with a red tie. "This is attorney Estevez from the US Attorney General's office," said Ted. I could practically feel my blood pressure[2] spiking at the mention of the top legal office in the country. I'd had no prior dealings with the AG's office up till now and was wondering what the hell they were doing here. But Ted had called the meeting, so I had to hope that everyone had come in peace.

But still, anxiety was my dominant emotion. Three major departments of government, all here to see me? Had they been listening in on my private conversations? Watching me? Seeing everything I did? What did they know about me? Had my actions put Donna, my mom, and my brother at risk?

Bottles of spring water were doled out to attendees as cell phones were silenced. I found myself gripping the armrests of my antique ball-and-claw chair as if the chair were of the electric variety.

Ted kicked off the meeting in his weedy Texas drawl. "I want to thank you for coming this morning," he said. "The reason I asked y'all here is

---

2. I have mentioned my blood pressure several times. My doctor now has me on blood pressure medication—a seemingly permanent consequence of that frantic period of my life.

that you represent three corners of the US government; three corners that should be coming together in a perfect and peaceful triangle. But right now, that triangle isn't quite fitting together.

"Since my client made the courageous decision to step forward and report the wrongdoings of his powerful international employer, his life has taken a decided turn for the worse. Not only has he put his career at risk and fallen into substantial debt to pay his legal fees, but now he's hearing that some of you want to hold him personally liable for his company's actions. Really? When he is doing all he can as a citizen to help you?"

I could feel Ted moving his gaze to each of the players, one by one. "The reason I asked y'all here is to gain some assurances that you don't plan to make my client's life any more difficult than it already is."

The US Attorney took a quick look at the others, as if seeking authority to speak for all of them, then slid to the front of his chair and cleared his throat. After a pause, he said, "We have no interest in prosecuting Mr. Yablon, or putting undue pressure on him. We appreciate his help, and all we are looking for—all of us—is his continued cooperation. This case, as you can imagine, has garnered attention at the very highest level of government. The very highest level. We would just like some assurances that your client won't develop a sudden case of cold feet."

Ted nodded thoughtfully. "I hear you, I do. But seriously, Ashley has met with y'all two or three times (looking towards Caldwell and Baxter) and has even given you his laptop to scan. How much deeper can he go? Let's be real, folks: he couldn't get cold feet even if he wanted to—because you guys have him sinking in quicksand." Everyone smiled. Not quite a laugh, but a bit of a thawing.

Ted went on, "Let me give you the elevator version of the story—just to be sure we're all on the same page." Heads nodded. "As you know, Mr. Yablon is employed by ZTE USA. A number of weeks ago, he was on a consulting and fact-finding trip to the company's headquarters in Shenzhen, China, when he came upon some extremely damning documents. These documents showed, quite unequivocally, that his company was illegally selling spying technology containing US-made components to Iran and other banned nations, in deliberate and flagrant defiance of US embargo law.

"Mr. Yablon struggled mightily over what he should do, as an attorney and as a man. He knew ZTE was in the wrong. But he also knew that if he ratted them out, he would be making an enemy of one of the largest corporations in the world and potentially ending his career just as it was getting started . . . "

Ted's words began to lose all sense of meaning, becoming echoey and distorted in my mind. It was as if he were referring to someone else, a character in a movie perhaps. Not me. Not Ashley Yablon of Dallas, Texas. I hate when people refer to themselves in the third person, but in this instance it is warranted. To me, he truly was talking about someone else.

All their voices dissolved into white noise. I couldn't hear what was being said or by whom. I was in the abyss. It wasn't until Ted put his hand on my shoulder again that I snapped back to the present moment. The meeting was already wrapping up, and Ted was giving his closing arguments.

"So, we're all in agreement then? Y'all will not come after my client personally, or pressure him, threaten him, or otherwise make his life miserable. In return, my client agrees to continue to provide information and to help you in any way that he can."

There was a round of nods and affirmatives. Everyone seemed pleased.

I wasn't sure exactly what I had agreed to. I just wanted to get the hell out of there.

Return to the Smoking Gun

The pressure from the Commerce Department backed off me a bit at this time. I don't know whether this was because DD&M was taking over more of the Commerce-related business or because of Ted's "triangle" meeting—or both—but I felt a slight lessening of stress from that direction.

That's not to say I was relaxed. I was still waiting for the next shoe to drop—for the FBI to pull its raid on the ZTE USA offices. I knew it was coming, and that was a constant source of tension, but I was managing to soldier on with my "life as normal" charade.

Until July 7, of course. That was the day I described earlier, when the fire drill occurred and I received the call in the parking lot from the Smoking Gun reporter. Hearing him inform me that he had possession of my sealed affidavit and was planning to leak it to the world in a few days' time sent me into a blind panic and kicked off the most stressful few days of my life.

As you'll recall, I phoned Ted immediately, and he went to work trying to see if anyone in government could help him get an injunction or otherwise quash the release of the affidavit. As you'll also recall, Ted was not optimistic about his prospects of success. Nor was anyone else I spoke to. The consensus among friends and lawyers seemed to be, "You're fucked, Ashley. This is terrible, this is unfair, this is outrageous, but nothing can be done."

Donna, meanwhile, was in Colorado dealing with her sister's cancer struggle. But as soon as she heard what was happening to me, she began making plans to come home. Waiting for her to get back to Dallas, and then waiting, the two of us, for the day the Smoking Gun piece was set to drop—hoping against hope that Ted would find a way to engineer an eleventh hour reprieve—felt like waiting for my heart to explode in my chest. Those few days probably shortened my life by five years. I'm not even kidding. I knew if that article and affidavit came out, I was a dead man, plain and simple. But I had no power to affect anyone's actions.

And then, of course, came that crazy day Donna and I spent lying on the floor of our bedroom, staring at my laptop, hitting the refresh button over and over as Gable hovered over us. The Smoking Gun piece came out late that day, July 12. Donna told me we had thirty minutes to get out of the house, and we hopped into the car to make our getaway. And things really took off from there.

CHAPTER 14

Code Red

So that's where I'll pick the story up now, with Donna and I sitting in her car outside our house with our hundred-pound Bernese Mountain Dog and our travel bags of thrown-together clothes and toiletries. Once again, we felt as if we were in a movie.

We needed to escape. We needed to get away from our home, and we needed to do so immediately. But where to go and what to do?

"Let's find a hotel," said Donna, who was at the wheel—I was too keyed-up to drive. "We'll check in under my name." Donna, luckily, had kept her maiden name Yarborough to use in her legal practice after we got married, and she still had credit cards under that name. "Then we'll figure things out from there."

Right. A hotel then. We needed someplace to be, a place to get our bearings and figure out our next move.

Maybe we were already too late. A BMW with tinted windows came rolling around the street corner behind us. Donna surveyed the car in her side mirror and shifted into drive. She drove at casual speed to the next intersection, turned the corner slowly, then took off down University Street at breakneck speed. She wove through traffic at twice the speed limit for several blocks. The BMW did not appear to be following us as far as I could tell.

No sooner had we departed our house than my cell phone started blowing up. When I finally pulled it out of my back pocket ten minutes later, I saw nearly two dozen missed calls and voicemails. I threw the phone down on the floorboard, but the calls kept pouring in, one after another. They were all from news outlets. I didn't want to take their live calls, but I did listen to some of the voicemails. "Hello, Mr. Yablon, this is so-and-so from CNN." "Hi, I'm looking for Ashley Yablon; this is so-and-so from the New York Times." "CNBC is interested in an exclusive interview." And so on. ABC, Newsweek, MSNBC, Google News, the Huffington Post—every news organization you can think of. The avalanche of calls was nonstop.

Donna drove frantically around Dallas as we looked for a low-profile hotel or motel we could check into. We quickly discovered that most hotels don't want to rent you a room when you have a dog the size of a St. Bernard in your guest party.

We finally found a room at the pet-friendly Hotel Palomar. We pulled into the valet parking lot so our car would not be seen. I trotted Gable to a small patch of grass to do his business while Donna dashed inside to check us in. I grabbed our bags, met Donna inside, and we took the elevator to our room, hiding our faces like escaped convicts.

The first thing we did upon entering our room was draw the shades, and then, as if on cue, plop down on the floor with Gable. Neither of us needed to explain to the other the reason why: to reduce our rifle-target profile.

We had no idea if our home had been bugged, or for how long, or whether a GPS tracker had been placed on our car. We didn't know whether we'd been followed here or if we were being watched. Nothing was a known. Certainty had been thrown out the window.

All we knew for sure was that I had just been outed on a national stage as the guy who had cost his Chinese employer billions of dollars in lost business. And when billions of dollars are in play, "professionals" are often hired to clean up the mess. We were genuinely afraid of being murdered.

As we sat on the carpet with our backs against the bed, each with a hand on Gable and the other on the floor for balance, my phone continued to ring every few seconds.

"Turn that goddamn ringer off!" Donna whisper-yelled. I did.

After several deep breaths, I gave Ted a call.

"I assume you saw the piece?" I said to him.

"I did. Where are you now?"

"We're in a room at the Hotel Palomar close to your office. We're huddled on the floor, hoping we don't get shot." Panic laced my voice. "I'm dead, Ted, dead."

"Hang tight. I'll make some calls and get right back to you."

I sat there watching my phone as it rang silently, over and over—continuous calls from magazines and news networks.

About ten or fifteen minutes later, Ted managed to wedge in a call amongst all the reporters. I quickly tapped Answer.

"Okay, Ashley," he said. "I called the FBI. They're putting together an emergency meeting, and they'll get back to me shortly with the specifics, so be expecting my call on that."

"Okay. I will."

"I also called a physician associate of mine. He's written you a doctor's note explaining that you've come down with something and will be out of work for a few days. I'll have him scan it and email it to you. That'll buy you a little time with your employer."

"So what are we supposed to do now, Donna and I?"

There was no answer.

The heavy hitters

Donna and I spent the night sitting on the floor of that darkened hotel room, dodging calls from the media, sneaking outdoors every couple of hours to take Gable for short walks. It was as much fun as it sounds.

Solitary confinement reached a tipping point after a day or two, and a dog the size of Gable needs a yard. So Donna and I decided, what the hell, to take our chances and return to the house. It was hardly a safe choice, but we couldn't live in that stupid hotel room any longer. All three of us were climbing the walls.

We threw our things together and headed home. As we turned down our street, we scoped out the scene. No TV news vans camped on the street. No reporters on the lawn. Good. The bad news, of course, was that we were moving back to our well-known, publicly listed address. If someone wanted to do us harm, they would now know exactly where to find us. Feeling safe had become a thing of the past.

Stepping into our little bungalow, I felt a rush of warmth. I didn't realize how much I cherished the home Donna and I had made for ourselves until we were forced to stay away from it.

Just a few years earlier, Donna and I had used the $10,000 we won in the Korbel Perfect Proposal Contest not on the honeymoon Korbel had intended but on a down payment on this place. Our first night after getting the keys, we were painting the walls during a massive Texas thunderstorm. Lightning struck and we heard a massive crash. We rushed outside in the rain to see that our huge tree had been snapped at the base and was resting across our roof. Welcome to home ownership.

But we also did great home improvement projects there. Inspired by our time on the Bayou, we designed a New Orleans-style courtyard— decked out with recovered brick and lionhead wall fountains. We even hired the gentleman who had done the ironwork at the Dallas Zoo to design an old-school iron gate that made us feel like we were strolling through the Garden District. We often hosted our families in that courtyard for Easter and Mother's Day brunches—catered by the two of us. Those memories came flooding back as I thought about the possibility that we might be forced to vacate our house permanently.

Finally, Ted called back and told me he had a meeting set up with the FBI and the Commerce Department.

The next morning, as I put on my Uniform, I kept my focus on hope—hope that whatever plan Ted had put together, and whomever we would meet that day, would help me gain safe passage to my next stage in life. But fear was right below the surface. I mean, it's one thing to meet a couple of FBI agents at Ted's office; it's a whole different ballgame to go to the FBI's Dallas Field Office at One Justice Way—home to the storied agency that had handled the JFK assassination. In all my years in Dallas, I'd never set foot in the place.

I drove into the parking lot, spotted Ted's car, and pulled up next to it. Ted was sitting inside, making notes. I felt my intestines tighten as I grabbed my briefcase and locked my car door. As Ted and I strolled toward the glass doors, I felt like an actor trying to play a scene that hadn't been explained to me.

Once past the checkpoint (a famous country singer, bizarrely enough, went through before us, laughing and smiling), we were shown to a massive conference room that, once again, looked like a movie set. A huge table, four inches thick and the length of a bowling lane, ran up the center of the room. At the far end of this table hung the US flag on the right side, the FBI flag on the left, and a picture of Barack Obama in the middle. I was duly intimidated.

"Relax," said Ted, patting my shoulder, "Nothing to worry about. I've been to these offices a million times. Though never in a conference room this size."

Thanks, Ted, very reassuring.

We were waiting only a moment when, without preamble, the door swung open and a line of men and women in business suits came filing into the room, introducing themselves and handing me their business cards as they passed. There were twelve of them in total, all from the Commerce Department, and they took all the seats on one side of the table.

A moment later, another line of suits filed in. Twelve more people, this time from the FBI. Same drill, intros and business cards. They took all the seats on the opposite side of the table.

My God, twenty-four government agents. What was this, the 9/11 Commission? Finally, after everyone had taken their seats, a mountain of a man strode into the room. He walked up to me and said, "Hello, Mr. Yablon. Pleasure to meet you. I'm Agent X (I can't recall his name) of the FBI." As he reached out to shake my hand, I noticed a detail that made my pulse race: his pinky finger was missing. That tiny feature spoke more eloquently about who this man was and the types of experiences he had been through than any CIA dossier could possibly have done. He strode to the head of the table and took his seat there with deadly authority. I sat at the other end of the table, placing the tower of business cards next to my note pad.

Ted, as always, was unfazed. He opened the session with a statement similar to the one he'd offered at the "triangle" meeting but even more impassioned and urgent. "Ladies and gentlemen, my client Ashley Yablon is in a highly distraught state of mind. He fears for his life, and rightly so. He came to you people, at great personal and professional cost, with an offer to help his country by providing extremely sensitive information about his massive multinational employer. He did so voluntarily and under assurances of complete confidentiality. And now he has been exposed—completely compromised—through a careless or malign act that occurred under your watch. He will never again be hired in his legal field. He may soon be disbarred. And not only has he become toxic in his profession, but he and his wife fear that they may literally be killed. In short, you've ruined his life."

The digitally challenged mountain at the head of the table nodded thoughtfully and said, "We agree. We don't know how the leak happened, and all we can say is we're sorry. As for the danger Mr. Yablon is in, well, if this was the Mexican Zetas we were dealing with, your client would already be dead. The Russian mafia, they're probably the next most dangerous entity out there. The Chinese? I'd say they're probably third."

"Is that supposed to make me feel better?" I blurted out reflexively.

A round of light laughter went around the room.

"No, it's not," said Mount Pinky. "Your peril is real, Mr. Yablon. And we can do a few things for you. Option number one: we can offer witness protection for you and your wife."

What? Witness fucking protection? Was he serious? I'd hustled and sacrificed for a decade to land a General Counsel job at a major corporation.

Donna had worked equally long and hard to put herself in a position to start her own law firm. Our friends, family, and social life were all in Dallas. And now we were supposed to just disappear? To where? Juneau, Alaska? Dubuque, Iowa? Dye our hair and pull a Saul Goodman? Assume an alias and get a job at a Cinnabon in a shopping mall? That was to be our thanks for helping the government bring a case against a huge Chinese company it had been suspicious of for years? This was to be how my life plan unfolded? I couldn't believe my ears. Witness. Fucking. Protection.

"That's not really a viable option for us," I said, gritting my teeth to contain my outrage.

"Very well. Option number two is we can give you a special phone number. If you're ever feeling in jeopardy, you can call the number and we'll have agents to your location within three to four minutes. We'll also notify the nearest police station, and they will be there in equally short order."

I was thinking, *Those options both suck. I risked everything to blow the whistle, and this is the best my country can offer?* I was watching in real time as my life went up in smoke.

"I feel like I'm in jeopardy all the time," I explained to the giant man. "We just returned to our house after being on the run."

"Oh, that's one other thing we can do: sweep your house for bugs and other security issues. Would you like us to do that, Mr. Yablon?"

"Yes, please."

Ted chimed in. "These measures are fine for safeguarding his life, but what about his livelihood? His career is over; what are you going to do about that?"

Mount Pinky frowned in thought for a moment and replied, "Give us a minute, would you, gentlemen?" He gestured politely toward the door.

Ted and I left the room and went into an adjacent conference room that was empty.

"What the hell's going on?" I asked my lawyer. "Why did we have to leave?"

"They're probably going to come up with some kind of offer for you," said Ted. "I gotta tell you, Ashley, in the thirty years I've been coming to this office for meetings, I've never seen this many heavy hitters flown in from DC for one meeting."

I didn't know whether to be encouraged or terrified.

About ten minutes later, we were called back into the big room. We took our seats, looking down the two long rows of suits at Mount Pinky, who sat dead center in the vanishing point. "We have an offer for you," he said.

Ted and I waited.

"We're willing to pay you to help us in our investigation."

"Okay," Ted responded tentatively.

"Unfortunately, we can only do such an arrangement for three months, maximum."

"Fair enough," said Ted. "And what would my client's compensation be?"

"We can offer a thousand dollars."

A thousand dollars? Was I hearing correctly? That was a fraction of my living expenses for one month. I'm sure my face must have telegraphed my disbelief. I was about to tell Mount Pinky, in polite terms, where he could stick his thousand dollars, but then a wiser voice spoke up within me: *A thousand dollars is better than zero dollars.*

"Can you excuse me for a minute?" I said. "I need to make a call before I accept your offer."

I exited the room and went back to the empty conference room where Ted and I had waited. I called Harold Gregory, my employment lawyer.

"You can't accept that deal!" Harold told me with alarm in his voice. "If you take a new job, even for a thousand dollars, you'll be waiving your employment claims against ZTE." Oh. Good thing I called him.

I walked awkwardly back into the meeting room, twenty-six pairs of eyes staring at me. "Unfortunately, on the advice of my employment attorney, I can't take that deal."

"All right, then," said Mount Pinky, with a wrapping-up tone. "We'll go with the FBI emergency phone number and the local police alert for now, and we'll sweep your house for bugs." He stood up, and the two rows of suits rose in synchrony on either side of the table. "Please give some more thought to witness protection, Mr. Yablon."

His words washed over me like someone else's reality. I felt numb.

As Ted and I collected our belongings and walked out of the conference room in a daze, we were stopped by Agent Baxter at the elevator. "Ashley, can I trouble you for another minute?"

"Sure. What do you need?"

I put my briefcase down as she opened a manila folder filled with 8x10 black-and-white photographs. She spread them across a nearby table.

"Can you identify these people?"

Nearly all of them were Chinese. Some I knew from ZTE—such as Guong One, Guong Two, and a few others—while a couple I didn't. As I made my identifications of the faces in the photos, I felt, once again, like a character in a spy movie.

"Thank you, Ashley, we'll be in touch," she said as she hurriedly collected the photos.

As she walked off, Ted and I made our way down the elevator and out of the building.

"You look like you could use a drink," said Ted when we reached the parking lot.

"You are an observant man." I replied.

You want me to what?

I never heard back from the FBI as to if and when they made the sweep of our house or what they found. Was the place bugged? Had someone broken into our home and planted listening devices? Had they done so after we'd left for the Hotel Palomar or before? How long before? Days?

Weeks? Months? Who had done it? What might they have overheard? What did my bosses know? How screwed were we? No answers.

Maybe nothing had been found. But I had my suspicions. From that moment forward, Donna and I never discussed anything sensitive in that house again. Conversations took place in the backyard with the sprinklers going.

When I returned from my big meeting at the FBI office, I barely had time to put my briefcase down or answer Donna's question about how things went when my phone rang. It was Harold Gregory, my employment lawyer, again.

"Ashley, I need to tell you something and you're not going to like it."

Harold, the master of bedside manner. I braced myself. Today had already been a disaster. What could possibly be next?

"I've been thinking about how best to preserve your employment claims against ZTE," Harold said. "It occurs to me that, as of now, you're still employed by the company; no one has made any move to terminate you."

"Correct. So?"

"That means you need to go back to work on Monday morning."

CHAPTER 15

The Belly of the Beast

"He wants you to what?" said Donna, her face freezing in shock.

"Go back to work."

She stared at me as if I was speaking Mandarin. "You're not going to do it, right? I mean, you can't. Not after what's gone down."

"According to Harold, I don't have a choice. Not if I want to retain any leverage in this situation. At this point, they haven't fired me, or even threatened to. So I can't give them a valid reason to do so. That's why we had to get a doctor's note to cover my absence for the last few days. Harold says I need to play this strictly by the book—show up, do my job, put the burden on them to terminate me. So when they do fire me, or try to drive me out, it will be a clear case of retaliation."

"It's not worth it," she said with a decisive shake of her head. "You need to stay the hell away from there. That place is toxic."

"Donna, the employment angle may end up being the only viable strategy we have against these people. Otherwise, you and I may end up with nothing but a pile of six-figure lawyers' bills. Believe me, I don't like this any more than you do. But I have to go back to work. We've got no other choice."

"No. Uh-uh. Walking back into that building would be suicidal."

"Tell me something I don't know."

We stared at each other in silence.

The tension was broken when a text message pinged its arrival on my cell phone. It was from Harold Gregory. "Dr. Molina put in a call to the HR department at ZTE. He told them you'll be back to work Monday morning."

Shit.

Crossing the return threshold

That weekend was quite possibly the most anxious one of my life. Barely a minute passed that I wasn't worrying about going back to ZTE. Donna and I tried to distract ourselves by doing anything we could: going out to dinner, catching a movie, taking Gable on long walks. No use. My brain kept compulsively returning to thoughts of Monday morning the way a tongue returns to a painful tooth. I had no idea what was in store for me there, and the scenarios I kept playing out over and over in my mind became increasingly wilder.

Monday morning, I was up before the crack of dawn. I hadn't been able to sleep a wink anyway. My hands were shaking as I buttoned my shirt and tied my tie. The notion of actually showing up at my job was certifiably insane. I had just cost my employer billions of dollars, and now I was going to stroll into the office and act as if nothing had happened? "Good morning, Meghan, how was your weekend? Xi, do anything fun and exciting?"

Ridiculous. But if Harold was right, I had to play this out.

The drive to work that morning seemed to take forever, and yet it was over in a flash. I was hyperaware of every turn I took, every second that passed at every red light. With each tap of the gas pedal, I could feel the endgame drawing nearer.

I parked the car in my usual space and stared up at the office floors that I once imagined would be my gateway to a whole new realm of success and opportunity. Now here I was, trying to muster the strength to just step in the door.

While sitting there in my car that morning, it struck me that nothing had gone as I'd hoped it would at ZTE, right from day one. My dreams had been pure illusion, and now I felt like a failure. I had always thought that by working hard to become a GC, I was setting Donna and myself up for success. And now look at where life had brought me: a pathetic fool, sitting in his car in a parking lot, psyching himself up to walk back into the belly of the beast in the vague hope of safeguarding some esoteric employment claims against his employer. Hardly where I'd seen myself years ago. I had started out my career like a house on fire. And today the final timbers of my illusions were about to crash and burn to the ground.

Oh well. So much for self-reflection. It was time to do what I'd come here for.

I texted the word "arrived" into my phone, as instructed by the FBI. A moment later, I saw a call come in from "Blocked." I knew who it was, of course. Agent Baxter.

I tapped Answer.

"Good morning, Ashley. I see you made it. Okay, I know we've been over this a couple of times, but I just want to run through it again. You are not alone this morning. We have eyes on you. There are four plainclothes officers stationed around the ZTE parking lot. They will be watching everyone and everything that enters and exits the building. If they see anything threatening or suspicious, they will move in immediately.

"In addition, you have the special phone number we gave you. Use it immediately if things go off script up there. Good luck."

I looked around the parking lot. No agents in evidence. But I guess that was the point. The presence of the FBI did little to comfort me anyway. I wasn't expecting to get shot in the parking lot (though I didn't rule out that possibility entirely). No, I suspected that if ZTE meant to do me harm, they would do so in such a way, and in such a place, that the FBI wasn't going to be of any help.

I exited my car and started my feet moving toward the building entrance. I felt as if I was taking a long, slow walk toward my own execution. My heart was beating so loudly in my chest, my footsteps automatically fell into rhythm with it.

I walked up the stone steps and opened the familiar glass entrance doors. Several people were waiting for the elevator; some made eye contact, some even smiled at me. Not all were ZTE employees, but somehow I convinced myself that all of them knew.

Stop it, Ashley. Jesus.

As I rode up in the elevator, waiting as it stopped at a couple of floors before my own, I wondered if my security badge would even work anymore, or if it had been disabled. Part of me was hoping for the latter. If I was locked out of my floor, that would mean I had already been terminated. I could just turn around and go home.

I arrived at the tenth floor and stepped out of the elevator. As I approached the French doors with the large glass panes that served as the entry point to the ZTE offices, I saw the receptionist talking on her phone. The moment she caught sight of me, her eyes bulged out as if they were trying to pop from her head. She leapt up out of her seat, still clutching her phone. I could hear her through the door shouting, "Oh my God, it's him!"

I stepped up to the doors and waved my security badge at the lock. The little LED light turned green, and the electric bolt clicked open. My badge still worked. Crap. The doors opened and I stepped inside.

Welcome back

The next hundred seconds were the most surreal ones of my life. It's a cliché to say that time stood still, but it did. Or rather, it slowed to a crawl. Each microsecond passed like a minute, and even now when I recall the scene, I can see it only in slow motion—like a car accident.

As I passed through the doorway, the phone literally fell from the receptionist's hand and her jaw dropped open. She stared at me in abject horror as I crossed the foyer and stepped into the main office area. I chose not to make eye contact with her. I was on a mission.

As I mentioned earlier, the bulk of the large office space was a cubicle farm, with the offices of the executives arranged around the perimeter. My office, of course, was at the far end of the room. Which meant I had to walk past the entire sea of cubicles, and all the other offices, to reach it.

The moment I rounded the receptionist's corner and entered the big room, all of the Chinese employees rose slowly from their seats, on cue, and faced me in silence. It was like a climactic scene from *Norma Rae* or *Dead Poets Society*. Everyone acting in synch, no one saying a word.

I continued walking—holding my head up, clutching my briefcase, eyes fixed on that distant wall where my office was located; an actor performing a role, refusing to betray what was going on beneath his stony veneer. The Chinese employees all just stood there, staring at me in deathly silence as I crossed the room—which now seemed the length of twelve football fields. I felt like a dead man walking—which I believe was the intent of the participants.

Every face glared at me as if I were a child killer. Every step I took became progressively more difficult.

As I got closer to the back of the room, something began drawing my eyes to my office door: crisscrossed yellow Police Crime Scene tape. I stopped and stood in front of the door for several seconds, staring at the yellow tape, unsure what to do.

Fumbling in my pocket for my keys, I glanced across the hall to the office of Meghan, Zhang, and Matthew. Its door had been removed from its hinges. My legal staff's office now stood empty; all of its furniture—its décor, its whiteboards, everything had been removed, all except for one desk, which was now pulled up close to the empty doorframe.

The desk, I soon realized, had been placed there to gain a perfect vantage point into my office, which was catty-corner across the way. Seated at the desk was a Chinese man I'd never seen before: early forties, heavy-set, black suit, black shirt, black tie. On his desktop was nothing. No phone. No computer. No tablet or pen. He just sat there staring at me, arms folded and wordless, like a military sentry.

Under the relentless watch of sentry-man and the gaze of all the standing Chinese employees, I reached up and tore down the crime scene tape. It made a loud and violent ripping sound. Keeping my face neutral, I unlocked the door and stepped inside. *Take your intimidation and shove it.*

My office had clearly been picked over in my absence. I noticed my laptop was gone, as was my office phone. As I turned to take inventory of whatever else might be missing, I found myself facing the huge whiteboard that occupied most of my office's western wall. The sight I saw there caused my knees to buckle and air to rush into my lungs.

Printed across the whiteboard, in bright red marker, in letters three feet tall, was a single word, marked with three exclamation points:

DIE!!!

I felt my stomach clench.

Was this some form of Chinese reverse psychology? My ZTE bosses had asked me to be their scapegoat on Capitol Hill, to swear under oath to their falsehoods. And now they were acting as if I was the one who had committed the crime by telling the truth? As if I was the one in the wrong for whistleblowing their illegal schemes? As if I was the evildoer in this thing?

Looking back on that day now, as I sit writing these words nine years later a new perspective begins to emerge on the whole scenario. Yes, it is true that the crime scene tape, the DIE!!! message, and the "security guard" watching my every move had been put in place to intimidate me; that's for sure. But the message went deeper, I now believe. It was also intended for the rest of the office staff, especially the eighty percent who were Chinese nationals—those "996ers" who had sacrificed so much to be here in America, away from their families, living in cramped apartments with their fellow workers.

The message was: We will not tolerate anyone who speaks up against the company or against the Republic of China (which, as I've said, dictates the actions of virtually every major Chinese company). Dissenters will be punished.

Looking back now, I wonder if the display the workers put on that day—standing up in unison as they did—actually might have been an acknowledgment of what I had done, fighting back against the bully. Again, I'm not painting myself as a hero, but maybe their standing in solidarity had been a way of saying *we appreciate you fighting the silent giant under whose thumb we all operate; we appreciate you calling out ZTE, calling out our country, taking a stand.* It sure didn't feel like it at the time, but now I wonder.

I was still standing there gaping at the whiteboard when someone stepped through the door behind me, causing my startle reflex to fire. It was Xi's secretary.

"Xi would like to see you in his office," she said to me.

"Why?" I replied, ridiculously.

She didn't bother to answer, simply shot me a freighted glance.

I grabbed a notepad and pen and followed her toward the CEO's office. This was going to be tons of fun.

As soon as I stepped inside Xi's office, he said, "Shut the door, please."

I complied.

"Why would you do such a thing?" he said with his trademark furrowed brow, wasting no time getting down to business. "Why did you make up these lies?"

"I can't talk about this, Xi, and you know it."

"Why, Ashley? Why the lies? Why? Why?"

I felt the urge to defend myself, to assert that every word I'd spoken to the FBI was the absolute truth, but I knew better than to engage with him. I was a lawyer, after all.

"If you have questions about any of that stuff, you can talk to my attorney," I said. "I'm here to work, so if you have company business to discuss, let's get started."

"No. I want to know WHY you did what you did, Ashley," he continued in a louder voice, anger fairly steaming from his pores. This was the Xi I knew and had braced for. "Has this company not treated you well?"

I wanted to reply, "not particularly," but again I resisted the urge. I could see what was going on here. The gaslighting campaign had started in earnest. Apparently, the new rule was that whenever not threatening me or pretending I didn't exist, everyone was supposed to act like I was crazy or lying. *I'd better brace myself for more of the same*, I thought.

It occurred to me that Xi might be recording this conversation; I had to make sure I didn't say anything that could be used against me in any way.

"Until I hear otherwise, I will assume I still have a job here," I said. "And so that's what I'm going to do, my job. That will be a little tough without a staff, by the way. Where are Meghan, Matthew, and Zhang?"

"They have been reassigned."

"Why?"

"Not your concern."

"Well, I'll need all their files, all their emails, and all the docs they were working on. Speaking of which, I have no computer anymore. So I'll need a computer to work on, too, right away."

"I'm sorry, can't afford one."

"Excuse me?"

"Can't afford one. Not in the budget."

"Xi, I need a computer to work on. You're not telling me, with a straight face, that a company with $700 billion in revenue can't afford a computer for its General Counsel."

"Yes, that is what I am telling you."

Unbelievable. So that was the game they were going to play.

"My phone is gone, too," I said. "Can you at least replace my phone?"

"Sorry. Can't afford that, either."

"Come on, Xi, don't be ridiculous. I need to communicate with people. I'm juggling dozens of projects and issues right now. We have litigations pending, contracts being drafted . . . "

"Sorry. Not in company's budget."

"For the last time, Xi, are you going to provide me with a computer and a phone?"

Silence ensued. It seemed destined to go on forever.

Finally, I stood up, walked to the door, turned back to Xi, and said, "Excuse me, then, I've got a job to do and I need to go do it."

Someone talk to me

I couldn't believe it. This was how they were going to treat me? I needed to talk to someone, anyone, ASAP. Storming past my office, I caught the glare of my new sentry-man from his post across the hall. I began sticking my head into offices and conference rooms, looking for any familiar and sympathetic face. Each Chinese employee I encountered looked right through me as if I weren't there.

Had there been a training session in my absence? Had a memo gone out? *Ignore Ashley, pretend he doesn't exist.* Was everyone in on the ghosting?

I finally spotted an American gentleman with whom I was casually friendly. He took a nervous look around to see if anyone was watching him before giving me a nod to enter his office.

"Matt," I said, "You've got to talk to me."

He looked around again and motioned for me to shut his office door.

"What's going on?" I asked him. "Where did my team disappear to? Why the hell is everyone freezing me out?"

"Seriously, dude? You have to ask that?"

Of course I didn't. "Fair enough," I said. "But what happened to my office? Who sacked it? Where's my computer and my phone?"

"You haven't heard?" He lowered his voice. "The FBI raided us on Friday, man."

"What? Holy crap. Xi must have been shitting himself."

"That's the weird thing, Ashley. Xi didn't show up that day. Neither did any of the Chinese nationals. Not a single one of them. In the whole company."

"What? That's, like, eighty percent of our work force," I said.

"Tell me about it. This place was a ghost town. Someone obviously tipped somebody off. And word went out amongst the 'favored' employees, if you know what I mean. The only ones who showed up for work that day were us dumb-ass Americans."

So the FBI raid had happened while I was out, and every Chinese employee had known not to show up. Baxter and Caldwell had warned me that my computer and phone system might be seized as part of the raid. I realized it was the FBI that had taken my things, not my bosses. As for my staff, I had no idea what had happened to them (and I never found out or set eyes on them again).

The new "normal"

I went back to work—with no staff, no computer, and no office phone. My cell phone was my only means of connection to the rest of the company and to the outside world.

Coworkers literally pretended I did not exist. When I would walk down the hall, people would look right through me as if I weren't there. The only Chinese employee who acknowledged my presence at all was my security guard who remained posted in the doorway of Meghan, Matthew, and Zhang's old office. He was there when I arrived each morning and there when I left each evening. His form of acknowledgement was to stare directly at me, literally eight hours a day. That was his full-time job. He did not speak to me, and I chose not to interact with him. Nor did anyone else. I never saw him speak to anyone at ZTE, not even once. He just sat at his desk each day, arms folded, watching me.

I received no internal phone calls or emails during that time. None. *Fine, so be it* was my attitude. I still officially had a job, and so I was going to do it to the best of my ability, despite all the obstacles my employers were throwing at me. If they were trying to smoke me out and force me to quit, they were going to have a long battle on their hands. I am nothing if not a patient man.

My cell phone was my only office tool, so that was my only way to send and receive documents, and in those days you couldn't print directly from a phone. Luckily, I managed to find one American guy in the sales department who would let me print emails from his workstation. So the way I had to work was this: When I would receive, for example, a draft of a contract by email, I would email the document to this guy in sales. He would print up the contract for me. I would then take a red pen and, in 1940s fashion, mark up the document by hand with my edits. I would then scan the paper document in his office, email it back to myself, and then email the redlined version to whoever needed to see it. The model of efficiency.

And so it went. I made it a point to come in to work a little earlier or later than most of my coworkers in order to minimize the chances of running into anyone. Every day at lunchtime, I would go out to a local restaurant alone or sit in my car and eat lunch by myself.

It was a lonely and weird existence. True Chinese torture. As uncomfortable as things were at work, though, they were even worse on the home front.

CHAPTER 16

Scenes from a B Movie

For the next several weeks, my life and Donna's became a bad spy movie—but one that was terrifyingly real. Our everyday life went something like this:

Each morning I would walk the dog, avoiding all pedestrians by zigzagging off the sidewalk and taking mental notes on any stranger I spotted more than once. Each time I'd cross the street, I'd scan for nearby cars that might be camped out to watch me.

After my dog-walk, I would drive to work, checking in the rearview mirror all the way to make sure I wasn't being followed. On more than one occasion, I was pretty sure I did have a tail on me and took one of several detours I had planned out ahead of time. There was one three-day period when the same car—I swear it was the same one—started tailgating me in the fast lane of the highway, causing my heart rate to soar.

When I would arrive at work, the routine I described earlier would kick in. I would make my way to my office, avoiding the incoming crowd as much as possible. All day long I would try to work under the watchful eye of sentry-man. Each time I left my office, either to go print out my emails in the sales office or to sneak out to eat a solitary lunch, I would be studiously ignored by everyone I encountered, to the point of absurdity.

When I'd come home in the evening, the first thing I would do was drive around the block, checking for suspicious parked cars, then do a walk-around of the house to make sure there were no signs of break-in or surveillance.

In the evenings and on weekends, if Donna and I wanted to talk to each other about anything sensitive, we maintained the practice I mentioned before. We'd go out in the back yard, turn on the sprinklers, and speak in a whisper. Even though our house had supposedly been swept of bugs, we had no confidence it hadn't been broken into again.

And then there were the clandestine meetings with the FBI in the Lincoln Town Car.

Confidential Human Source (CHS)

Not long after my big meeting at the FBI office, I received a call from Agent Baxter while I was sitting at home watching the early evening news. "There's something we need to run by you, Mr. Yablon."

"Okay," I replied. "Do you want me to come to your office tomorrow?"

"No, we'll come to you. Watch out your front window. An agent will pull up in front of your house. When you see his car, exit the house, and get in the car with him."

Sure enough, five minutes later, exactly, a black Lincoln Town Car with tinted windows came rolling up the street. It inched its way past our property line, paused for several seconds, then slowly backed up and parked in the evening shade of our tree.

I threw on some shoes, exited the house, and casually strolled toward the car. As I was drawing up alongside it, the passenger door opened and I slipped inside.

An agent dressed in a black suit and black tie pulled out his credentials and showed them to me. After we exchanged polite greetings, he opened a briefcase and plucked out a large manila envelope. Unclasping it, he withdrew a stack of black-and-white photographs as Agent Baxter had done a few weeks earlier. He handed me each photo, asking if I could identify the individual in the picture. Some faces I knew, others I didn't, as before. Once again, I felt like a character in a spy movie.

This little ritual in the Town Car—which would repeat itself several more times over the course of the next year or so—was carefully staged to send a message about the FBI: powerful, omnipresent, not to be fucked with. Message received.

This was a truly crazy period in our lives. Andrew Silver, my employment lawyer in New Jersey, only reinforced the madness when he called me one day and said, "Hey Ashley, I was thinking about your case, and it struck me that it's a pretty big story. A General Counsel for one of the biggest corporations in the world suing the company *while he's still working for them.* Would it be okay with you if we worked with the firm's PR company and published a piece about it? The story and press would really help us out."

"Sure, knock yourself out," was my basic reply. I had become oddly detached by this point. I had bigger fish to fry. Such as staying alive.

Death threats

One weekend day, Donna and I were out shopping just to get out of the house. We couldn't afford to buy anything, and we were certainly not in "acquisition" mode, what with the possibility of going into the witness protection program looming over us at any moment. But sitting around the house was making us feel like ducks in a pond surrounded by hunters. So we would go out as often as we felt we could.

We were at a store about fifteen miles from home, trying to amuse ourselves, when a text came in on my phone from Meghan. This was weird. First of all, Meghan had been reassigned and no longer worked for me. Second, the call was from Meghan's old number. Several months earlier, Meghan had received a new company-issued phone. She had turned the old phone in to the company and had been assigned a new number. When I'd set up her new contact in my phone, instead of deleting or updating the old contact I had simply created a new contact called "Meghan 2." Why, I don't know.

But now I was being contacted by her old phone, which I knew was in ZTE's possession.

The text itself was the really alarming part.

I no longer have the actual text, but some things you don't forget. It read something like: *We are going to kill you. We know where you live, we know where you go, we know what you do, and we are watching you. We will kill you and make you suffer. We will kill your wife, we will kill your family members.*

The blood rushed from my head, and I felt as if a carpet had been yanked out from under my feet. Looking around for watching eyes, I pulled Donna out of the store. We got into the car and drove to a nearby restaurant to have a drink and try to calm our nerves. As we sat there, a few more texts of the same nature came in.

The texts accomplished their goal of scaring the shit out of us. We didn't feel safe anywhere now. Not at home, not away. I forwarded the threatening texts to the FBI, but in typical fashion, they did not share any further information back with me. I get it, they're in the information collection, not the information sharing business. But it certainly would have helped my paranoid psyche to be given some assurances that everything was going to be okay.

Or maybe everything wasn't going to be okay.

That was the frustrating part.

Yellow cab

Another day—this was after receiving the death threats—Donna went out to walk Gable. As she was heading down our pathway to the sidewalk, she noticed an odd sight. Parked in front of our house but at the top of the street (we lived on a corner) was an old-school yellow taxicab, the type with the 1960s-style body design. It was a jarring sight to see on our tree-lined suburban street where we seldom saw cabs of any kind, but especially not such ancient New York City dinosaurs. Donna immediately recalled a Stephen King story in which bad guys from an

186

alternate dimension traveled in old yellow cars. It was not a comforting thought.

Donna took a right and started walking down our street. She had traveled only thirty yards or so when she turned her head to see the cab slowly rolling down the street after her, keeping a steady twenty-yard distance. Donna slowed, waiting for the cab to accelerate and pass her, but it didn't. It hung back.

Donna felt her pulse pick up. She hurried ahead to the corner and turned right, hoping the cab would just move on. After she'd made the turn, the cab turned too and moved up closer, which gave her a chance to look at the driver.

He was a middle-aged Chinese gentleman.

Normally, the racial/ethnic identity of a driver would be irrelevant, but in this case, it was a hard factor to ignore.

Donna's heart began to pound; she stepped up her walking pace. The cab increased its speed to stay exactly in step with her. She fumbled in her pockets for her cell phone, but all she could find was a wad of doggy clean-up bags. Donna never leaves home without her phone, but on this day she apparently had. She broke into a half-trot, turning another corner to continue around the block. The cab stayed with her.

Donna still thought there was a possibility the cab might be harmless. When she got to the next corner, she turned right, and now began jogging along the side of our house. With the cab at her heels, she picked up her speed even more. She ran around the final corner onto our street where the cab had originally been parked.

She was in full sprint as she headed up the walkway to the front door. The cab paused, then sped off down the street, too fast to get the license plate number.

Meghan?

Then there was the time I was at work and received a text message from Meghan 2. That was Meghan's new number, so it grabbed my attention.

The text read: *Ashley, why are you telling these lies?*

I stared at the screen, anger building, trying to decide what the hell my response should be. Do nothing? Reply and try to engage her? Instinct told me not to reply, just to ignore the text. But part of me was worried about my old friend. If this was the real Meghan texting me, and she was really writing such words, then she was probably doing so under threat or duress.

Against my better judgment, I texted her back: *Meghan, is that you?*

The reply—and again I don't have a record of this exchange or a precise memory of the words—was another generic *Why are you lying?*

I texted back: *If this is you, Meghan, then you know I am not lying. You know what's been going on at the company. You and I have talked about it many times.*

Her answer: *Why are you saying you saw things you didn't see? Why?*

So the gaslighting campaign was in full bloom. Trying to make me out to be an unreliable witness. Trying to make me out to be nuts. Trying to make me doubt my own sanity, my own memory of events and documents.

I wondered again what had happened to Meghan. Had she been fired? Harmed? Was this really Meghan texting me or was someone within the company using her phone? Was she being actively threatened and coerced into texting these words?

I texted another message, which I hoped would trigger an honest response from Meghan: *I am not lying. You know what was said and what they are doing.*

There was a long pause, and then a new text appeared: *Why the lies, Ashley?*

I knew at that point it was not Meghan on the other end. I was dealing with a ZTE operative trying to wage psychological warfare. The "magic" was gone. For good.

I shut off my phone.

Close encounter

One evening, Donna and I decided to go out to dinner. We didn't like to dine near our house; we could never relax for fear of being watched. So we went to a restaurant about twenty or thirty minutes from our home.

The place was almost empty. We walked in and took a booth. We were studying the menus and enjoying our drinks when two Chinese men strode into the place. They could have sat anywhere, but they chose the four-top directly across from us, literally a few feet from our table.

They sat on their chairs, and when the waitress brought them menus, they pointedly lay them face-down on the table and pushed them aside. They then turned their chairs so that they were facing directly toward us, folded their hands in their laps, and proceeded to sit there staring openly at us.

"Time to use the bat phone?" said Donna through her teeth, trying not to move her lips. The bat phone was our way of referring to the special number the FBI had given us.

"I believe so," I whispered back, pretending to read the menu.

As nonchalantly as possible, I slipped my phone out of my pocket and flipped through a couple of screens, as if looking for a text or an app. I found the FBI number and hit the Call button. Holding the phone a foot away from my mouth, still pretending to read the menu, I said, "I think we need some help," as if talking to Donna about deciding on a meal.

We continued this dining charade for a couple more minutes as the two Chinese men continued to stare at us.

Within four minutes, I spotted a man and a woman in suits approaching the entrance door from outside. FBI, no doubt about it. I signaled Donna. We stood up, I threw some cash on the table for the drinks, and we marched toward the door. As soon as our feet began moving, the Chinese men stood up and followed right behind us. I didn't have to say a word to the agents as they entered; they knew what to do. Donna and I just hauled ass to our car, jumped in, and sped off, never looking back.

So that was life for us in the post-Smoking Gun weeks.

Another interesting shift that was happening was in my relationship with DD&M. Whereas before the Smoking Gun story and my filing of the lawsuit, the law firm and I had always been on the same side, at least theoretically—things were now getting murkier. DD&M now had a much stronger motivation to paint me as the crazy outsider. I had also learned that they would be representing ZTE in my lawsuit against the company. So we were now in this weird position where I had to keep working with them in order to fulfill the duties of my job as General Counsel of ZTE, while at the same time I had to navigate the adversarial relationship that was developing between us. Walking that line was tricky.

One morning, for example, I was on my way to work when I received a call from Jeff Pennington. He put me on speaker phone with one of his partners. They were asking me questions ostensibly related to the House Intelligence investigation, but it was clear they were really probing me for information. At one point in the conversation, the second attorney slipped and said the word "affidavit," and Jeff quickly corrected him, saying "subpoena." I knew that the conversation was a fishing expedition—DD&M was trying to find out how much I'd told the FBI about them.

Meanwhile, my own lawyers' fee-meters were running and I was continuing to borrow money to feed those meters. I was striving to keep communications open and honest with everyone I owed money to. Everyone was extremely generous and non-demanding, but my steadily deepening debt was adding enormous stress to an already insanely stressful situation.

More than once I felt the stress threatening to make me snap. I would ask Donna to step outside, we'd turn on the sprinkler system, and I would say to her, "Am I nuts? Is all this stuff really happening? Are we really being followed and threatened? Did I really see what I thought I saw over there in China? Everyone is saying I'm crazy. Am I? Did I do the right thing? Should I have just kept my mouth shut?"

Paranoia strikes deep, as the song goes.

Donna always said the right thing, telling me I wasn't crazy and that this stuff truly was happening to us. She was unfailingly encouraging. But I knew this whole debacle was starting to affect her too. We had burned through all of our savings to support me at a time when she really needed my support—dealing with her sister's cancer and starting her own private practice. The collapse of our dream was troubling her deeply. She didn't want to say it aloud, but the tension was present and mounting. And it was causing stress cracks to form in our marriage.

CHAPTER 17

Leave

On the morning of August 20, 2012, exactly thirty days after my return to work at ZTE, Donna and I were sitting in our kitchen, sipping our usual caffeinated beverages. I was dreading my workday and regaling her with my litany of complaints against my employers: the shunning, the constant live monitoring as if I were a prisoner on suicide watch, the refusal to provide me the tools for doing my job (computer and phone). "The Chinese invented water torture," I mused, "and it's as if they're using a mental form of it on me—this steady drip, drip, drip of abuses—in hopes they can drive me to quit on my own."

Donna turned to me, paused with her can of Sugar-Free Red Bull halfway to her mouth, and uttered the understatement to end all understatements.

"Ashley, you are in a hostile work environment."

We looked at each other for a second, then burst out laughing—that blackly comic kind of laughter that had become all too common for us in those days. It was the absurd obviousness of the comment that made it so funny. We were cry-laughing as we traded follow-up commentary.

> **Me:** "I don't think death threats are all that bad, do you? I mean who doesn't have 'DIE!!!' written across their office whiteboard once in a while, right?"

> **Donna:** "For sure. Show me an employer who doesn't occasionally resort to having their employees' wives followed by stalkers in yellow cabs! I challenge you!"

On and on it went. Finally, after we'd caught our breath, she said, "I'm serious, though. Your situation at ZTE clearly exceeds any legal threshold for a hostile work environment, and you can't passively tolerate it any longer. Here's what you need to do. Before you go to work today, you need to sit down at your computer and write a letter detailing your work conditions and make a formal complaint against ZTE. I mean it. You need to do that right now, and you need to take that letter to the head of your HR department and hand-deliver it. Today. It's time to go on record. You need to establish some claims against these people."

She was right. For some reason, the simple idea of filing a complaint had never dawned on me, no matter how bad things had gotten at work. Maybe it was because I was so used to being on the other side of the fence—the attorney defending an employer against hostile work environment claims that often amounted to little more than a boss raising his or her voice. Never had I contemplated being on the receiving end of the abuse—and never to this extent. Or maybe I had just become like the battered spouse who starts to passively accept their partner's cruel acts. I was so lost at that point I couldn't tell.

But I did know this: when Donna has a strong instinct, I don't argue with it. And she was right that I needed to write a letter to ZTE.

I went to her computer and drafted a letter. As a General Counsel, I had handled employment-related matters and litigated employees' claims against the company, so I knew the right buzzwords to use and the right aspects of my work environment to call out. In my letter I asserted that my supervisors' actions were pervasive, severe, and disruptive, and that they made it impossible for me to do my job effectively. I stated that ZTE was aware of the actions and behaviors of its employees toward me and had done nothing to intervene. I also stated that I'd been denied the tools to perform my job. I emphasized that the actions of my bosses were having a detrimental effect on my career and that they "altered the terms, conditions, and/or reasonable expectations of a comfortable work environment for employees," and so on.

The words flowed easily onto the page; after all, I'd had a whole month to stew about them. It was a damn good letter. I printed it, put it in a white envelope, thanked Donna for her advice, kissed her goodbye, and drove off to face another day in hell.

It turned out Donna had been prescient. Shortly after I stepped into my office that morning, an administrative assistant from HR popped her head in the door and told me the HR department wanted to see me. Ah. So perhaps ZTE was finally ready to call an end to this ridiculous charade.

When I arrived in the Human Resources office, the head of the department, Glen Summers, and a Chinese executive were seated waiting for me. Glen was a pleasant enough fellow. An older gentleman with white hair, Glen had been working for ZTE USA for some time. I knew him well. Together we had handled many employee issues over the previous ten months—so I was keenly aware of ZTE's risk tolerance for hostile work environment claims.

I sat down, and Glen started in. "Well, Ashley, considering everything that's been going on at the company . . . "

"Wait," I interrupted. I knew where this was going. "Before you say anything, I have this letter I would like to officially submit."

I slid Glen my white envelope, and he accepted it, but he never opened the envelope nor asked what was inside. Without missing a beat, he continued, informing me that the company was placing me on administrative leave. He handed me a large manila envelope containing the relevant paperwork. "You'll continue to receive a salary until your status with the company has been resolved and we notify you otherwise. Do you want to collect your belongings now, or shall we arrange to have them sent to you?"

"I'll take them now," I replied. I handed over my security badge and the keys to my office. Under the watchful eyes of an HR representative who followed me to my office, I gathered my belongings. Much like that morning thirty days earlier when everyone had stood in silence upon my arrival, my exit was excruciatingly awkward. I felt like Bud Fox in *Wall Street* being led out of the office by the feds to a silent chorus of stares and glares. Here was the company's chief legal officer, doing the perp-walk with his belongings in a cardboard box.

It was the last time I set foot on ZTE property.

On leave

It might sound like a dream to be put on administrative leave—collecting a salary without having to show up at work. And I'll admit, for the first week or two it did feel like a relief. Anything was better than having to work under the conditions I'd endured for the previous thirty days. But administrative leave gets old fast. I wasn't able to look for new work, and I couldn't resign from ZTE, because if I did I'd be waiving my claims against them. I was in limbo, career-wise. And that was not where I wanted to be. At forty years old, I was in the prime of my working life, and I wanted to be doing something meaningful. I wanted to be contributing to the world and advancing my career. I didn't want to be stuck in some weird semi-retirement holding pattern.

I did have my pending cases—my lawsuit and employment claims against ZTE—to work on, so I spent a good part of my wide-open schedule doing legal research, talking to lawyer friends, and trying to map out a strategy on those fronts. But there was only so much time I could devote to those tasks without risking my mental and emotional health. I've seen people who become virtually obsessed with righting wrongs they believe have been done to them. They focus on their injury night and day, year upon year, until it consumes them mentally and emotionally. They end up doing more damage to themselves via their obsession than was ever done to them by way of the original offense. I didn't want to be one of those people. True, my wrongs had been particularly nasty ones—but I wanted to keep some balance in my life.

People often ask me why I didn't just quit and move on. My answer is that it wasn't that simple. I had bills to pay, and, again, Donna was still trying to build her private practice. We both had college loans and law school loans to pay. And besides, getting a job as a General Counsel would be no easy task. And I'd just made it quadruply more difficult by becoming embroiled in a highly publicized adversarial situation with my current employer. No one wants to hire a GC who is known as a problem-maker rather than a problem-solver.

Plus there was the fact that I was still a central cog in the whole ZTE investigation. I was the government's star witness. And the case hadn't played out yet. No matter how much I might try to distance myself from ZTE and/or downplay my role in the whole scandal the fact was that I was still deeply involved. And there was no avoiding that reality or hiding it from a new employer. In truth, there was no practical way to move on from ZTE at the moment.

Also, my lawyers' bills were continuing to stack up. They were already in the six figures by this time. If I walked away from my job, the chances of my collecting a financial award from ZTE would plunge precipitously, or so my attorneys kept telling me. No money award from ZTE meant no money to pay my attorneys or pay back the people I'd borrowed from. Which meant I would have to pay everyone back over time out of any future salaries I might earn. Which in turn meant prolonged financial hardship for Donna and me.

No, there was no choice. I had to play this thing out. I couldn't just cut my losses; I had to double down on my bet that I could collect some form of damages from ZTE. Which meant I had to keep collecting my salary and keep the lawyer train rolling.

Chugging away

I had to keep the lawyer train rolling for another reason as well: the criminal implications. I was still in potential jeopardy on that front. As the government's case against ZTE continued to heat up, I knew ZTE and

DD&M would try to scapegoat me as much as possible. And when one of the wealthiest telecom companies in the world and one of the most powerful law firms in the country team up against you, you don't dare leave yourself unprotected. So Ted Masters, as my criminal attorney, continued to be an essential resource.

I had good reason to worry about what DD&M might try to do to me. Why? Because, frankly, I'd witnessed them doing things that made me question their ethics. I didn't know how they might behave when feeling threatened. For their part, I think they were nervous about me too. After all, I'd seen the inner workings of ZTE. I'd also seen some of the inner workings of DD&M and had been vocal in my disapproval of their handling of certain ethical and legal issues. I think they were worried about what I might have on them.

Specifically, they wanted to know what I had said to the feds. When I had still been actively employed in my GC role, they had tried to coax this information out of me in various indirect ways. Now that I was on administrative leave, they were forced to come after me more directly. They told me they wanted to depose me.

Well, they didn't want to call it a deposition. You see, a deposition is taken under oath; you have to swear to the truthfulness of what you say. And every word is recorded by a court reporter who then transcribes it into written form. Every question. Every answer. Under such a scenario, DD&M would get to ask questions, but so would Ted. The court reporter would later certify the entire deposition by a sworn affidavit. A deposition has the exact same weight as a witness on the stand at a trial, so a deposition is a big deal.

DD&M didn't want that kind of permanent record—they wanted to call our session an informal Q&A instead. Why? Because DD&M was nervous about what I knew, what I had told the feds. And if any of the information I was holding turned out to be damaging to DD&M, and if it was then recorded and sworn to with an affidavit, well then . . . that could be a problem. DD&M is one of the top lobbying firms in the US. If their reputation were to be tainted by my words, many people would go down. As Donna likes to say to her clients, "This could be a bad one."

In fact, it was Donna who insisted that the informal Q&A session be recorded by a court reporter. It had to be official, she felt. (But of course, court reporters aren't cheap. Hiring one would mean another $1,500 out of our pocket plus Ted's time to sit there and defend me.)

DD&M's position was that the session should not be recorded. Their reasoning: *Ashley, you're still employed by ZTE, even though you're on administrative leave. Talking to us informally is your duty as an employee. You are required to participate in any company investigation*—and that was what they deemed this to be, a company investigation.

We went back and forth with each other. I finally acceded to their request but with a major stipulation. We would have a court reporter on hand to capture Ted's argument as to why the Q&A session should be recorded—and DD&M's argument as to why it should not—but we would not record the actual questions and answers. My thought was that by taking at least this tactical step we would preserve my justification of my claims. DD&M didn't like it, but they finally agreed.

This was to be a simple question-and-answer session of which no form of record was to be created; nothing I could turn around and try to use against them. We met in Ted's office and did the Q&A there. Basically, I told them the same things I had told the FBI. When they tried to pry from me what I had said specifically about DD&M, I told them the truth. The fact was I hadn't said much to the FBI about DD&M—partly out of professional courtesy and partly because the FBI's focus had been almost exclusively on ZTE. But that didn't mean I couldn't say anything about them in the future. I tried to convey to DD&M, in a subtle way, that I still remembered some of the questionable things I'd seen them do, and that I might perhaps know a few additional things as well. I didn't threaten them in any outright way, but at the same time I didn't want them to walk away from the deposition thinking, *Ashley's told us everything; we don't have to worry about him anymore.*

I wanted them to worry. I thought it might give me some bargaining power.

I continued to consult with my informal legal team as well—my lawyer friends. They were in many ways more helpful than the lawyers I had on payroll.

Several months passed without making any progress in my claims against ZTE, and I began to conclude that I'd gone as far as I was going to go with Harold Gregory. I had preserved my employment claims by going back to work (even if that hadn't been necessary), and I was now on administrative leave. There was nothing more for an employment attorney to do at this point. Things overall were at a standstill. But I wanted to move them along. I wanted resolution.

It was around that time I began to think, "I'm actually a whistleblower. Maybe that's the angle I should be approaching this from." I had learned of a federal program that allowed government whistleblowers to be paid a percentage of the money the government was able to recover as a result of their whistleblowing. I'd read of cases where whistleblowers were paid literally tens, even hundreds, of millions of dollars. It occurred to me that maybe the kind of attorney I really needed was a whistleblower attorney.

I started looking into the matter and was surprised to learn that one of the country's top whistleblower attorneys was actually located right in Dallas. His name was Shawn Kellerman. On his website I learned that he had written a book on whistleblowing, which contained a whole chapter devoted to blowing the whistle on Chinese companies. I called his office, and his assistant informed me I would need to pay a consulting fee, a pretty hefty one, just to sit down with Shawn. I couldn't afford it. But at that point my attitude was *I'm already paying four attorneys I can't afford, what's one more?* I made an appointment.

A couple of days later I drove to Shawn Kellerman's practice in downtown Dallas. As soon as I stepped into his office, we both froze and drew back our heads.

"I know you," he said.

"I know you," I replied.

After a moment, we both realized why. Shawn was my across-the-street neighbor. We often waved to each other when grabbing our mail or when coming home from work at the same time.

Shawn is a white-haired giant of a man. Quiet and seemingly standoffish, he's actually quite warm and personable. Over the years I'd seen him many times playing with his black lab, Buster, in his front yard.

"Did my assistant tell you there'd be a charge for today's meeting? Don't worry about that; no charge."

I liked the guy already.

I explained my current situation to him and told him my story in a condensed way, ending with a disclaimer that I really couldn't afford another attorney.

Shawn made a generous proposal. He offered to handle my case for a one-time flat fee of $7,500. I believe he made the offer from a genuine place of good will, but he was also interested in my case professionally. He felt it could contribute to a newly revised edition of his book and would be good for his portfolio. I humbly asked if I could pay him in installments. He agreed.

"The first thing we need to do, Ashley, is manage your expectations. You don't technically qualify as a whistleblower, at least in terms of those whistleblower reward programs." My heart deflated a little. He explained that those programs were reserved for people who worked for the government or government-contracted companies, which I did not. But he did say there were other statutes in place that might benefit me.

"Maybe with your knowledge of whistleblower statutes," I suggested, "you can put some pressure on ZTE to come to a settlement with me."

"That'll be my goal," he said.

I gave him the green light and hitched another car onto the lawyer train.

Mediation

More months passed with no real forward motion on either front—my lawsuit against ZTE to pay the legal fees in my defense against the Commerce department or my administrative leave issue. *Seriously, ZTE: how long are you going to keep me on administrative leave?* With each passing month, my impatience to achieve a resolution grew. I was pushing Shawn Kellerman to try to get ZTE to settle both issues, and he was doing his best to pressure them.

Finally, in April 2013, in an attempt to both settle the lawsuit and obtain a fair severance deal from ZTE, I forced a mediation. Here's where Chad and Donna were proven right. It was my lawsuit and my "hostile work environment" letter that gave me the leverage to bring ZTE to the mediation table. Without those two actions, ZTE never would have talked to me.

Mediation is an attempt, prior to a trial, to get the parties to negotiate and reach a settlement. The mediator is a neutral person, usually an attorney or former judge, who plays the role of good cop/bad cop and negotiates with both parties (who remain in separate rooms during the proceeding). Imagine a car salesman running back and forth between you, the buyer, and a supposed "manager." He keeps telling you he really went to bat for you to get you the best deal. Sure, okay. The mediator is doing that in both rooms.

It is said that a good mediation settlement is one in which neither party ends up feeling like they really won. In other words, both sides need to give in order for a settlement to cross the finish line.

I asked all my lawyers to attend the mediation session. This required, in addition to paying for their time and expenses, flying Andrew Silver in from New Jersey, putting him up in a hotel, forking out $2,500 for the mediator, and taking the whole team out to dinner the evening before the mediation. Donna and I thought this was the right thing to do in spite of the cost. If all went well in mediation, I would soon be paying them all their final balances and saying my goodbyes.

We took them to a high-end Dallas restaurant called Stampede 66. As the first round of appetizers was placed on the table, I stood up and raised my glass to the team: Andrew, Shawn, Ted, and, of course, Donna. "I want to thank you all for being here. We've all been through quite a journey to get to this point. I deeply appreciate all the concessions you've all made for me, financially and professionally. Here's to a resolution tomorrow and to getting this ZTE debacle behind us once and for all."

As Donna and I drove home that evening, we talked hopefully about get our lives back at long last. We prayed ZTE's counsel would be reasonable. But we were holding our collective breath, knowing all the stars would need to be perfectly aligned for this thing to settle.

The next morning, we all met at the mediator's office. ZTE was represented, of course, by DD&M—not the lobbying specialists like Pennington, but employment law partners flown in from Washington, the same ones who had done my Q&A session a month earlier.

If you've never experienced mediation, it's an all-day affair. It usually kicks off with what's called a joint session—it should be called a bitch session—in which each side gets to air out its grievances against the other side. As an attorney, I've been involved in several mediations over the years, and it's my opinion that the joint session does more harm than good. All it does is fire up antagonism between the parties and make each side hunker more deeply into its entrenched positions. I advocated for skipping the joint session and jumping straight to negotiations.

My plea fell on deaf ears.

After the joint session—which went exactly as I feared it would—the parties retreated to separate rooms for what is known as private caucuses. That's when each side meets privately with the mediator, who then goes back and forth, discussing the relative merits of each side's case and presenting offers from the other side. The car salesman routine.

In our case, it was obvious from the beginning that our mediator didn't really get my side of the argument or seem to care. It was just another five-thousand-dollar day to him ($2,500 from each side). He kept saying things to me like, "I don't understand. Can't you just get another job?"

I tried to explain to him why that wasn't so simple. After several unsuccessful attempts to get him to see my perspective, I lost my temper. "You're a god-damn lawyer. What the hell would you do in my position?"

"I'd do what we all do," he shrugged. "Ask around. Go to my network. Get a job through friends."

I wanted to explode.

"I'd still have to find a company that would hire me!" I said. "ZTE has ruined my reputation. They've made me into a liar and a scapegoat! I'll be lucky if I don't get disbarred. And what do I have to show for it? Do you think ZTE is going to give me a glowing reference for my resumé? There's no way I'm going to find another General Counsel job right now! I'm fucking unhireable."

"So how do we settle this?" he sighed, arms folded and toes tapping. "What do you want from them?"

I was astonished to realize I hadn't come up with a numerical figure for my damages. I looked to my lawyers—who should have done this for me—and they pretty much shrugged.

In all the mediations I had ever done for clients, I always had a dollar figure in mind as to what my client's damages were. I sometimes even hired "damages experts" (think accountants and insurance actuaries) who used mathematical calculations, charts, graphs, and historical data to show just how much my clients were entitled to receive. It's a curious thing: I was always able to advocate for my clients and tell them exactly what they needed to do. But when I became the client, simple things like determining my own damages escaped me.

I quickly did some rough calculations in my head. I started with my salary for a year, factored in how long it would likely take me to find a new job, threw in some exemplary damages.

"I don't know. A million dollars?" I said.

I looked to Ted Masters and he shrugged and said, "Sure, why not?" He was working on another case on his laptop.

The mediator rolled his eyes, took my offer to the other room, and came back a while later with the generous counteroffer from DD&M of $0.

I then offered $750,000.

They counteroffered $0.

I offered $625,000.

They countered $0.

"This is ridiculous, guys!" I shouted at my team. "I give them a number, they counter with zero. I reduce my number, they stay at zero. I'm fucking negotiating against myself!"

Finally I asked Shawn to go out into the hallway and explain Sabine Pilot to the attorneys on the other side and why it applied in my case. Sabine Pilot (short for *Sabine Pilot Serv. v. Hauck*) is the seminal legal case establishing that a company is not allowed to retaliate in any way against an employee who points out that the company is doing something illegal.

Shawn summarized our main legal points: Ashley discovered that ZTE was buying US component-parts, illegally putting them into spying technology, and then selling this technology to the embargoed country of Iran. ZTE then asked him to lie to Congress, and when he refused, ZTE retaliated against him by refusing to pay his legal fees, threatening his life, having him followed, thwarting his ability to do his job, creating a hostile work environment, et cetera.

Shawn's intervention served to energize the discussions, and it seemed for a time that a reasonable settlement might be imminent. But when 6 p.m. rolled around— the end of a very long day—we still had no offer in hand. Finally, the mediator came back into the room and gave us the astonishing news that DD&M's lead attorney said she was not familiar with Sabine Pilot or its significance. That's like a PhD student in physics telling you they're not familiar with Sir Isaac Newton. Sabine Pilot is *the case* law books cover when you're a first-year law student. And now the head employment attorney at one of the biggest law firms in the US was claiming not to be familiar with it? If this wasn't retaliation against me, then nothing ever could be. In fact, I felt my case could now replace Sabine Pilot as the landmark study on employer retaliation going forward.

Realizing we were spinning our wheels, we finally left the mediation having made no progress. I drove home, thoroughly dejected and dispirited, and spent the evening wallowing in self-doubt. *What the hell am I up to? Am I doing the right thing? Why did I start this whole mess? Why didn't I just keep my mouth shut? How am I going to pay all these lawyers? What am I going to do with the rest of my life?*

If there is such a thing as the "dark night of the soul," I experienced it that night. Those twelve hours were perhaps my bleakest hours in the entire ordeal.

By morning, however, a flicker of hope had inexplicably returned. It probably wasn't justified, but it was there. And it was just enough to stop me from throwing in the towel.

CHAPTER 18

Reconciliation

And now we wind our way toward the end of this tale. I should warn you that, unlike many of the scenes you've read thus far, the ending does not play like a Hollywood movie. Real life refuses to follow a three-act structure. I wish I could tell you that after a disappointing mediation, we rallied and staged a dramatic courtroom showdown complete with stunning witness-stand testimony and a gigantic jury award that broke ZTE's back. Or that the leadership of ZTE was perp-walked out of its offices by the FBI on live TV with helicopters hovering overhead for additional news footage. Those things did not happen, sadly.

But this story was far from over.

Payoff

In the months that followed the mediation session, I continued to urge my attorneys—Shawn Kellerman mainly—to push ZTE for settlement. He tried, but ZTE still did not want to budge. My employers were seriously pissed at me and were resolute that I would not walk away with a dime. I was equally pissed at them and determined not to be left high and dry.

In the end, I think it was two dynamics that moved ZTE's hand. One was that we began to pressure ZTE (via DD&M) to submit to depositions in regard to the New Jersey lawsuit. Just as DD&M hadn't wanted to put my deposition on the record, they didn't want to see the ZTE brass fielding official queries from all my attorneys. Can you imagine the questions those guys would have been asked? Not only about their retaliatory actions against me, but also about their criminal behavior on an international stage? Worse, can you imagine their reaction to being placed on a witness stand in front of the world? Again, as Donna likes to say, "That would be a bad one."

We also began asking for written discovery—the documents ZTE planned to use against us in the case—as well as sending them interrogatories. These are questions each side is required to answer for the other side as a case draws near its court date. Essentially, we were asking ZTE to spell out—in writing, testimony, and documentary evidence—its specific reasons for refusing to pay my attorneys' fees, for resorting to retaliatory behavior, et cetera. And what became very clear was that ZTE didn't want to be pinned down on these points (because they didn't have any good reasons for their behavior and they damn well knew it).

Meanwhile, Shawn Kellerman was calling and emailing continually Claudia Mulroney, DD&M's lead attorney handling my case, pressuring her to settle. "Look," he was saying, essentially, "this has gone on long enough. How long are you all going to keep paying Ashley not to work? It's hurting ZTE and it's hurting him. You can't hire a new GC; he can't go out and look for a new job. Everyone's hands are tied. This is a lose-lose. Can't we just settle and move on?"

No response from ZTE. And so we sat. And sat. Again, ZTE was trying to smoke me out, see if I would just go away.

Finally, one day Claudia called me and told me that, as of a specified date, ZTE was planning to stop paying my salary.

I pointed out that they couldn't do that. "Am I being fired?" I asked her. "If so, for what reason? And by the way, what was the finding of that supposed 'company investigation' I was part of?"

I also explained that because we were still in settlement mode, ZTE couldn't fire me—that would be retaliation for my calling them out for illegal acts. She argued that they could fire me. We went back and forth. Finally, in exasperation, she just blurted out, "What's it going to take to end this? What is your absolute bottom line?"

So there we were, in a negotiating stance at last. With all the lawyers I was paying, it was no small irony that the final resolving of my claims came down to me negotiating on the phone on my own behalf. That point was not lost on me as a long-denied grin crept its way onto my face.

But regardless of my newfound glee, I was in no mood for playing games. I quickly calculated what it was going to cost me to pay all my lawyers, pay back everyone I'd borrowed from, and pay off a few other debts I'd incurred as a result of the ZTE fiasco, and I gave her that honest, no-frills dollar amount. (I'm not legally permitted to disclose what that amount was or even if there was an amount at all.)

We went back and forth on various points, things I wanted, etc. She hemmed and hawed, but finally said, "Fine."

And so just like that, it was done. On August 14, 2013, a little more than a year after I'd been put on administrative leave, I reached a resolution with ZTE. I became a free man at last, no longer a ZTE employee.

Hallelujah.

Life in exile

And little did I realize what sort of battle life after ZTE would be. The day I was officially set free from ZTE, I began trolling the job boards and sending out resumes. It didn't take me long to realize that no one was replying to me. No one. I was radioactive. Naturally I had expected to encounter some speed bumps when trying to rejoin the workforce, given what had gone down at ZTE, but I hadn't expected to be in absolute exile. But that was the situation I found myself in. It soon became clear that my struggles in getting out of ZTE were only the beginning. My real challenge would be trying to get in somewhere else.

I sent out resumes every single day. I wasn't even searching for a General Counsel job necessarily; I knew finding one of those would be super-tough. I was looking for any kind of in-house position I might land with a technology, software, or telecom company. I was looking at associate counsel jobs. I was looking at law firms, nonprofits, small litigation companies—anything my resume qualified me for. No matter what the job was, though, I would never get a call back. I'd reply to job postings that described me to a T—"We're looking for someone who has served in an international telecom company, knows how to handle contracts, and has litigation experience"—and I wouldn't even receive even a courtesy response.

I ended up being out of work for about two years in total. During that time, I sent out literally hundreds of resumes. Not a single nibble, not a single call back, not a single interview.

The only firm that would hire me was my wife's. I ended up helping her on a few of her cases—sometimes doing research, sometimes serving as second chair for a trial. We did a couple of two-week trials related to guardianship issues (Donna's specialty), that sort of thing. Donna wasn't exactly throwing me a bone by hiring me; I did have litigation experience and I did add value to the cases I worked on, but neither of us believed this was the best possible use of my time and talents.

It was actually good that we did this work together, though, because it helped Donna and I heal some wounds. Our marriage had actually suffered a lot during the whole ZTE ordeal. I alluded to this earlier, and I don't really want to go into much detail about it. It's private stuff, and I don't want to put Donna through the pain of publicizing it. Suffice it to say that the pressure of the case and the financial strain we were going through created a lot of stress to which I did not always respond brilliantly. Donna and I had a lot of fights, and most of them were caused, at bottom, by the tension of the ZTE thing. I failed as a husband in some pretty substantial ways and even ended up separated from Donna for a time and living with my brother in the extra bedroom of his apartment.

In an attempt to resuscitate our marriage, we had to make some life changes. One of those was to sell our house, the bungalow on the tree-lined avenue. The house had acquired some bad memories as a consequence of the whole ZTE nightmare. It also seemed like an unsafe place for us. We felt that someone could too easily do us harm there—break the door down or shoot us through the window—and we were still afraid of that happening.

We needed a fresh start, a safe place, a place with no memories, a place that had nothing to do with the ZTE years. So we sold the house and bought a condo in a Dallas high-rise, a smaller place for us. It would be much harder for someone to target us there, we reasoned, and it would be an opportunity for us to start building some new memories.

A lot of renovation work was needed at the condo, and so overseeing that work was another job that occupied much of my time during my radioactive period.

Resurrection

I had a tough time moving past the whole ZTE experience, I'll admit. Psychologically, it kicked me to the ground in a pretty profound way. I had always thought I was a reasonably grounded and world-savvy guy, so it shocked me to realize how crudely and gracelessly I'd allowed myself to be manipulated by ZTE. It caused me to call my judgment and self-image into serious question. The struggle over whether I was doing the right thing by reporting ZTE to the FBI ate away at my confidence even further. And the fact that no one would touch my resume for nearly two years was the icing on the self-doubt cake.

"All you need is that next opportunity," Donna kept telling me. "Once you get your first new job, it'll be like a palate cleanser; you'll be able to forget about ZTE and turn the corner on a new life."

Donna ended up being right about that. That annoying mediator also ended up being right when he'd said, "Can't you just get a job through a friend like the rest of us do?"

Nearly two years after I had officially settled with ZTE, I got a call from a friend of mine who helped me land my next job. It was a GC role for a Dallas-based company, a position I had never thought was again possible for me after sitting on the sidelines for two years, steeping in self-doubt. An added benefit, the office was a two blocks from our new home. But what made it even more special was that Donna and I were able to start a new tradition, which we have kept up to this day even as I've transitioned to another job: we meet every workday for lunch.

Change needed to occur after ZTE, and a rededication to my family life needed to be a major part of that change. Life was finally turning around for us.

And it turned out there was an even more satisfying chapter yet to come.

CHAPTER 19

Vindication

One early weekday morning in March 2017, I was sipping coffee while putting on the Uniform and getting ready for work. Through the white noise of the television playing in the background, a news story grabbed my ear. I cocked my head, wondering if I had heard it correctly: *Did they just mention ZTE in a major storyline?* Although I had tried to purge my former employer from my memory, the mere mention of those three letters hooked my attention like a car accident scene on the freeway. Fumbling for the remote, I turned up the TV volume but had just missed the story. I ran to my computer and logged on to the *New York Times* to see this headline: "U.S. Fines ZTE of China $1.19 Billion for Breaching Sanctions."

Holy crap, had ZTE finally been busted?

I read on to discover that ZTE had agreed to plead guilty to breaking US sanctions and selling electronic equipment to Iran. According to the article:

> "ZTE engaged in an elaborate scheme to acquire U.S.-origin items, send the items to Iran and mask its involvement in those exports," said the acting assistant attorney general, Mary B. McCord. "The plea agreement alleges that the highest levels of management within the company approved the scheme."
>
> She added that ZTE repeatedly lied to and misled federal investigators, its own lawyers, and internal investigators.
>
> . . . "ZTE acknowledges the mistakes it made, takes responsibility for them and remains committed to positive change in the

company," said Zhao Xianming, chairman and chief executive of ZTE.

There it was in black and white, the truth I had paid so dearly to bring to light, the truth that had derailed my career for four years, the truth ZTE had accused me of being a crazy liar for asserting. Admitted to at last.

From the article—and from clicking around the Web—I learned that in addition to pleading guilty to conspiring to evade the US embargo, ZTE had agreed to plead guilty to making false representations to the US government and to obstructing justice. It fully admitted to hatching the elaborate shell-company scheme way back in 2010 (a year before I joined the company). The articles all made mention of me as the former GC of ZTE and government whistleblower. And the kicker? My gaslighting former bosses were directly implicated in the plot.

As I dug through all the online reports, I uncovered the truth I'd known deep down all along. It was my quiet, tea-sipping, sweater vest-wearing, philosophy professor-looking Chinese legal boss, Mr. Guong (One), who had concocted the entire scheme—along with the CEO of ZTE China, Shi Lirong. I'd always suspected that Guong One must have masterminded the plot, but I had missed the red flags that had been waved in my face at the time.

I reflected back on all those meetings with Mr. Guong: me asking if we had anything to worry about, him reassuring me that we were "clean" yet assailing me with questions about finding the right law firm. I was now convinced it was Guong who had personally orchestrated the entire nightmare scenario I had lived through. It was him all along. Every detail of those meetings with him—the sights, the sounds, the smells—came flooding back to me at once. I was suddenly right there in his office again, looking at his pleasantly smiling mask.

As punishment for its flagrant acts, ZTE was socked with the largest fine ever levied in US history: $1.2 billion in criminal and civil penalties. I stopped reading for a minute, pushed back my chair, and let that sink in. The largest fine in US history.

ZTE agreed to pay the fine, no contest. It also agreed to reprimand its executives, fire everyone above a certain executive level, and submit to a new reporting and monitoring system that the US would put in place. ZTE agreed to pay about $900 million of its historic fine up front; $300 million of it was to be suspended pending the company's successful completion of a seven-year probationary period. In other words, if they played nice for seven years, they could keep the $300 million. (In my opinion, that amount was a drop in the bucket for ZTE and unlikely to influence its behavior in the least.) ZTE was also expressly forbidden from publicly commenting on or denying the allegations.

And what did ZTE get in return? It was allowed to continue operating in the US. And we all know that's really good for business.

ZTE had actually been found guilty of its sanction-breaching crimes about a year earlier, in 2016. At that time US authorities had determined that ZTE had lied to the US when it stated it had stopped its practices of selling to banned countries back in 2012. (ZTE had, in fact, blithely continued those practices; it had simply changed its scheme for doing so.) In 2016, US authorities also determined that ZTE had lied about other matters and repeatedly thwarted US investigations. One of the ways ZTE had done this was to require that all employees involved in sales to Iran sign NDAs (Non-Disclosure Agreements). An NDA states that you cannot disclose your stated activities to anyone.

Another way ZTE got around the US's investigation was to set up a thirteen-member "contract data induction team" (CDIT) whose job was to scrub the ZTE databases daily of information related to any Iran business that had taken place after 2012. Why did they do that? Because as part of its investigation, the US government had hired a world-class data analytics and computer forensics company to do a deep dive into all of ZTE's computer servers. So ZTE's CDIT scrubbers spent their workdays cleaning all potentially damning evidence from the company's computer systems. And in conjunction with this clandestine effort, the company created an autodelete feature for the CDIT team members so that their emails were erased each night, a failsafe system on top of the scrubbing.

In 2016, ZTE had been placed on the Entity List, which was supposed to require any companies that wanted to export tech products to ZTE to obtain a special license. That provision was never enforced, however, and ZTE continued to skate and dodge major consequences. It was only when the US succeeded, a year later, in having ZTE removed from the Hong Kong stock exchange that ZTE finally threw its hands up in surrender. Within two days of that action, ZTE showed up at the bargaining table, ready and eager to plead guilty to everything and accept whatever fines the US decided to enforce.

I acquired a copy of the actual settlement ZTE signed in 2017, and when I read it, my jaw dropped. It was like reading a duplicate of my FBI affidavit. ZTE admitted, point by point, to virtually all of the allegations I had made, including the names of the players and the names and roles of all the phony shell companies. ZTE's attempts to gaslight me were at last revealed as lies for all the world to see.

Vindication (whatever that means) was mine at last. Sadly, $120 million was not. As I noted earlier, I did not technically qualify as a whistleblower under the current statutes. Had I qualified, I would have been paid at least ten percent of the government's settlement. That extra $120 million sure would have provided me some decent "walking-around money" (as my dad used to say), but instead I would have to take satisfaction from my public vindication and my private mental images of ZTE's leadership being marched out the door by security guards with their badminton trophies and Buddha statues in cardboard boxes.

A month or so after the story broke, all ZTE leadership and board members were forced to resign. But guess who got a promotion and became the new CEO of ZTE Mobile Device Worldwide? That's right, my former boss and CEO of ZTE USA, Xi Kai. I can only imagine whom he threw under the bus to gain that position.

Have I said this before? You can't make this stuff up.

ZTE lies again

About a year after ZTE had pled guilty to all those charges and accepted the largest fine in US penal history, the company was at it yet again, lying and deceiving.

The US government had requested—as part of ZTE's mandatory reporting requirements—a follow-up report on the disciplinary actions the company had agreed to take with its executive team. It turned out—surprise, surprise—that ZTE had lied about those disciplinary actions. Instead of firing a whole raft of executives as promised, they'd in fact fired only four people and given bonuses (not fines) to about thirty-five other top-level people.

In response, the Commerce Department's BIS (Bureau of Industry and Security) activated a denial order that was tied to ZTE's seven-year probation. This order officially placed ZTE on the so-called "Denied Persons" list. US companies would now be prohibited from selling ZTE any components whatsoever. This was a massive blow to ZTE, a company that relied on US components for at least twenty-five percent of its smartphone construction. These components included the Android software and processors made by Qualcomm. For ZTE to re-engineer its products to avoid using such critical US parts would be a huge and time-consuming effort. And even then, without access to Android software and other US technology that allowed ZTE phones to be part of US networks, ZTE could never really be competitive in the smartphone market again.

ZTE was essentially finished. In May 2018, it announced that it had suspended most of its operations, including manufacturing, and the sale of its stock. Its employees were placed on furlough. This was a massive blow not only to ZTE but to China itself, which regarded ZTE as an important player in its quest to establish China as a worldwide technology leader. It was also a blow to US–China relations, which were already on shaky ground.

ZTE was desperate to find a way to get the sanctions reversed.

If you read the newspapers, you know that President Trump came to view ZTE as a potential bargaining chip in trade negotiations with China. He made it known that he was open to lifting the sanctions on ZTE, presumably in return for concessions from China on some other major trade issues.

Bottom line: ZTE was given yet another chance. This time, it was whacked with a second billion-dollar fine, plus an additional $400 million in suspended penalties, to be placed in escrow, contingent upon its behavior for the next ten years. Together with the nearly $900 million it had already paid, this put its total penalties north of $2.5 billion. ZTE also agreed to fire its entire board and senior executive team within thirty days and to pay the costs of an ongoing team of compliance officers selected and appointed by the BIS. Mr. Guong, and others who had put together the elaborate plan to deceive US authorities were finally forced to exit the building. That group included Xi, whose short-lived tenure as CEO of ZTE Mobile was brought to a close as well.

Today, ZTE is once again back in business. Does this mean the world can now trust that it has actually curtailed its banned international activities? An optimist might say yes. With its new board, new CEO, new public attitude, and a new monitoring team in place, ZTE seems to be on the right track. Me, I'm in the "fool me twice, shame on me" camp. I've seen these Chinese telecom companies from the inside, and I have absolutely no reason to believe ZTE isn't diligently working on some new scheme to flout US law. When I think of ZTE, I think of a kid holding a football for a younger sibling to kick. ZTE is the older sibling, telling the US, "Just kick the football. I won't pull it away this time, I promise." The US is the younger sibling, teeing up to kick the air once again and fall on its ass, Charlie Brown-style.

But hey, that's just my opinion. I could be wrong.

I am so glad to be out of that toxic environment. I hate to think of what I might have become of me if I had stayed. Sometimes it feels like yesterday that all this stuff was happening. More often it seems like another lifetime.

Onward and upward

In the years that followed, I tried to downplay my ZTE experience. After all, I'd put in a mere nine months of dubious work as ZTE's General Counsel. But I'd done some really productive and positive work at all my other jobs—jobs where I'd never been involved with the FBI, had my life threatened, or appeared on the front page of The Smoking Gun. I reasoned that people wouldn't want to hire a GC who had been the center of a whirlwind of trouble; they would want to hire a GC who knew how to keep trouble away. So as you can imagine, I tried to steer conversations away from ZTE.

But I soon discovered that just wasn't possible. Sometimes, when you're involved in a high-profile incident, that becomes the thing that follows you forever, no matter how you try to distance yourself from it. You either accept it as part of your legacy and try to build on it in a positive way, or you let it define and destroy you.

As I write these words, I am working for a firm that provides outsourced General Counsel services to companies. I am able to take my expertise and serve in a consultant role, helping companies streamline their in-house legal departments. It's interesting work with a lot of variety and challenge.

These companies and projects I now deal with are as different as can be, but one thing remains absolutely constant from place to place.

Whether I'm on an interview for a new placement or meeting with clients to go through my relevant work experience, some version of the following conversation always occurs—and I do mean ALWAYS:

"Well, Mr. Yablon, your resume is impressive. Your work experience is incredibly broad. But I do have one final request. We sent your resume around to the team and everyone had the same question. Can you tell us about this thing with you and ZTE?"

I always smile, take a deep breath, and . . .

The End

Epilogue

Reflections

It was only after I'd been re-employed for a while and had regained some sense of clarity and self-confidence that I was able to begin reflecting on the whole ZTE experience with any kind of detachment. As I thought about the story, and as I told it to friends and associates at social and professional gatherings, certain questions came up over and over. These were questions I sometimes asked myself in the early hours of morning, and that other people always wanted answered when they heard about my experiences.

How did I let it happen?

There were many moments in the ZTE affair—hiring a criminal attorney, meeting with the FBI, seeing the "DIE!!!" message on my whiteboard, going on the lam with Donna, et cetera—when I found myself asking the universe, *How did it come to this? Why me?* It was only after the dust had settled that I finally realized the questions needed to be directed at the mirror instead. The real question was, how did I allow myself to be put in the situation I ended up in? To answer this question required some self-examination.

I guess the short answer would be hubris. And naiveté, as I said at the start. I allowed myself to be blinded, to a certain extent, by the power of my own desires. I had spent years preparing to land a job as General Counsel of a major corporation, and I wanted the job so badly I could taste it. When the opportunity finally arrived, it felt like destiny rolling out the red carpet for me. My ambition trumped my critical reasoning, and I chose not to see the warning signs as they popped up. This is a pretty common flaw, I think. I was talking recently, for example, to a friend who'd striven for years to break into Hollywood as a screenwriter. One day he got a phone call from a producer claiming to be madly in love with one of his scripts. The producer offered to pay him $850,000 for the screen rights. The writer so wanted to believe his ship had finally come in, he signed a contract right away. He later learned his script was being used as part of a scam to lure backers into investing in a phantom movie that was never going to be made. Red flags were all over the place, but the writer couldn't see them—his dream had blinded him.

It was the same kind of thing for me at ZTE. The red flags were right there in my face—the routine withholding of vital information from me, the ignoring of my legal advice, the stated hopes that I would "testify on the Hill"—but I couldn't see them because I was too in love with the notion of being GC for a huge telecom business. It never occurred to me that I was being set up. In my defense, this was a Chinese company, so I chalked up a lot of its odd behavior to cultural differences, but I should have woken up sooner.

I won't say I have become jaded in the post-ZTE days. I still have optimism and I still have confidence in my talents, but I do pay a lot more attention to red flags than I used to. If something doesn't pass the gut test, I try get to the bottom of it right away.

Do I have any regrets? If I had it to do over, would I make the same choices?

I came to doubt myself many times during the ZTE ordeal, but now, when I look back, I see there really was no choice but to do what I did and come forward. There was no real moral gray area. At the risk of sounding self-righteous, I had a duty, as an attorney, to call out the illegal behavior I was witnessing. And I had a duty as an American citizen to help protect my country against the abuses of a foreign power. To do otherwise would have been a form of treason. So the choice was simple. Not easy, but simple. And yes, I'd do it over again.

What did I learn from this experience?

I did learn some valuable truths, truths that have made me a wiser, more grateful, and more resilient person. I learned that doing the right thing is hard. And that it sometimes exacts an extremely high cost. Doing what's right is easy when there's no price to pay; it's damn fucking tough when the cost is your livelihood.

On the other hand, the cost of doing the wrong thing is sometimes higher in the long run. This is what I try to explain to the people who cautioned me to just keep my head down and not make any waves at ZTE. If I had gone along and played the role ZTE was grooming me to play, there's an excellent chance I would be in federal prison right now. So maybe doing the right thing is also the wisest play in the long run, despite its costs.

I learned who my true friends are, and I learned the value of those friendships. The help I received—in the form of advice, encouragement, introductions and references, money loans—has humbled me beyond belief. I will never take friendship for granted again as long as I live.

I also learned what a remarkable, courageous, and committed person Donna is. She was my rock throughout the whole misadventure, and if our marriage was ever put to the test, it was because of my failings, not hers. I learned that she really meant it when she said that stuff about for better or for worse, for richer, for poorer. I learned how lucky I am to be married to her.

I learned to be a lot more adaptable and nimble than I used to be. I learned how to live on less. I learned how to change gears on the fly and let go of my attachments to expectations. I learned how to live in the now, not in the future, and to how find joy in even the most stressful of situations. I learned to reinvent myself.

Has any good come out of this?

I like to believe I helped our country become a bit more secure. I like to believe I helped prevent some of our foreign adversaries from gaining access to spying technology, with which they might have done our nation harm and with which they might have committed human rights abuses against their own people. (Literally as I was writing this paragraph, the *Wall Street Journal* released an article about how Huawei has been selling surveillance equipment to African nations to help their leaders spy on citizens and stifle dissent.) I like to believe I helped sound a cautionary alarm in our nation's evolving relations with China. I like to believe I took a stand for justice, as corny as that may sound.

On a personal level, I can tell you that one extraordinary thing has emerged from this ordeal. That is, I found out what I'm made of, so to speak. And that has been a true gift. I think every one of us wonders, at one time or another, how we would behave if faced with a moral choice in which the stakes are incredibly high. We ask ourselves, *If I had everything to lose—including my livelihood and, potentially, my very life—would I do the right thing?* Few people, however, are actually put to the test.

I am grateful to say I was put to the test. And I am pleased to say I passed it. I lost a great deal in terms of career and finances, but I gained something priceless: self-respect. I learned that when push comes to shove, I will do the right thing, even if the price is staggeringly high. And that knowledge has made me a better person. I like and respect myself more than I once did. And that's something no one can ever take away from me.

Are there any lessons others can take from my story?

Well, one little thing I hope people learn is that not all lawyers are morally compromised sharks!

On a more serious note, I hope readers will heed the cautionary tale embedded in these pages. As I noted at the beginning of the book, China is the new eight-hundred-pound gorilla in the global business arena.

Chinese multinational companies are now proliferating like kudzu after a rainstorm. Nearly every industry is being affected by China, and that trend is only going to intensify. China is extending her reach all around the globe and is busily building infrastructure in dozens of countries so that new technological Silk Roads can be established, all leading to one destination: China. Sooner or later—probably sooner—everyone who is in business is going to have direct dealings with Chinese companies.

A wise person will be prepared for this. Doing business with China is not the same as doing business with the US or other countries in the West. While Chinese companies may don the outer trappings of Western companies and may use the same buzzwords and business concepts, on the inside they are different from ours in some fundamental ways. I'm not saying Chinese companies are evil or underhanded. Not at all. What I am saying is that their loyalty is to China and to Chinese values and interests, no matter how Western they may appear on the surface. Most, if not all, of the big Chinese multinational companies have close ties to the Chinese government, and their decision-making is based on what is good for China, not what is good for the US. They respect Chinese law more than international law or US law, and they tend to regard Western laws that get in their way as inconveniences to be worked around. They play their cards close to the vest and do not show their full hand to Westerners.

I realize these are broad statements—and I'm sure they don't apply universally—but they have proven true in my experience. And in the experiences of friends and associates.

What am I trying to say? Avoid working for or doing business with Chinese companies? No, you probably couldn't even if you wanted to. I'm saying proceed with caution. Don't assume they are following the same rules you are. Enter relationships with eyes wide open, and keep your eyes open. Don't take appearances at face value. When someone tells you, "Everything is fine," as I have learned, it probably isn't. Never assume you have seen all the cards that are being played. I certainly didn't see everyone's full hand in this story.

Learn as much as you can about Chinese culture, Chinese values, and the Chinese business mindset. Learn what people are really saying—in their words, in their gestures, in their silences. I have offered you my experiences to help give you a "CliffsNotes" version, but you'll have to be your own judge. Just keep this in mind: China is busily learning about US and Western culture so it can better compete on the world stage; that means you need to do your homework and learn about China too.

When it comes to Sino-American business dealings, trust isn't fully warranted yet, on either side of the fence. We may get to that place some-day, but we're not there yet. Not in my view. Not by a long shot. Don't succumb to the "magic."

About the Author

Yablon is living proof that, sometimes,
making the easiest decision is the hardest thing you'll ever do.

Ashley Yablon was born and raised in Dallas, Texas. After graduating summa cum laude from SMU with a degree in political science, Yablon fell in love with the idea of becoming a lawyer while working as a runner at a local firm. He pursued his dream, going on to graduate from Loyola New Orleans.

Upon graduation, Yablon began an ambitious career, swiftly moving up the ladder toward his ultimate goal of becoming general counsel to a major corporation. He believed his hard work had paid off in 2011, when he accepted a position as general counsel for the multi-billion-dollar Chinese tech behemoth ZTE at just forty years old.

Yablon's dream quickly became a nightmare when he found himself in the middle of the largest scandal to ever hit the tech industry. When faced with a dangerous choice between career and country, Yablon risked it all to stand up for what is right. His refusal to toe the line is an inspiration to all Americans. Yablon is living proof that sometimes, making the easiest decision is the hardest thing you'll ever do.

Yablon lives in Plano, Texas. He works as a consultant, advising law practitioners and corporations on domestic and international dealings, as well as optimizing their legal departments.